High Performance Computing Symposium 2012

(HPCS2012)

Journal of Physics: Conference Series Volume 385

Vancouver, British Columbia, Canada
1-3 May 2012

ISBN: 978-1-62276-346-7
ISSN: 1742-6588

Printed from e-media with permission by:

Curran Associates, Inc.
57 Morehouse Lane
Red Hook, NY 12571

proceedings.com

Some format issues inherent in the e-media version may also appear in this print version.

Printed by Curran Associates, Inc. (2013)

Institute of Physics
Dirac House, Temple Back
Bristol BS1 6BE UK

Phone: 44 1 17 929 7481
Fax: 44 1 17 920 0979

techtracking@iop.org

Additional copies of this publication are available from:

Curran Associates, Inc.
57 Morehouse Lane
Red Hook, NY 12571 USA
Phone: 845-758-0400
Fax: 845-758-2634
Email: curran@proceedings.com
Web: www.proceedings.com

TABLE OF CONTENTS

PREFACE

The High Performance Computing Symposium (HPCS) is a multi-disciplinary conference that is considered to be Canada's premier HPC forum. Each year, researchers, industry experts, vendors and HPC innovators in academia and industry gather to exchange new tools, techniques and results in and for HPC and computational research.

HPCS 2012 (http://2012.hpcs.ca/) was hosted by WestGrid/Compute Canada and held in Vancouver, BC on 1–3 May 2012 alongside the BCNET Annual Conference. More than 550 HPC experts and tech innovators attended three days of workshops, demos, panels and presentations.

We would like to thank all the sponsors, speakers, staff, and delegates for helping make HPCS 2012 a success. Also, thank you to the authors of these papers and to the reviewers for their support in the preparation of these proceedings.

HPCS 2013 will be held in Ottawa, ON. For more information, visit http://hpcs.ca/.

As a national organization, Compute Canada coordinates and promotes the use of HPC in Canadian research and works to ensure that Canadian researchers have the computational facilities and expert services necessary to advance scientific knowledge and innovation. WestGrid, one of Compute Canada's largest regional partners, is composed of 14 partner institutions spanning British Columbia, Alberta, Saskatchewan and Manitoba.

The HPCS 2012 Organizing Committee

Peer review statement

All papers published in this volume of *Journal of Physics: Conference Series* have been peer reviewed through processes administered by the proceedings Editors. Reviews were conducted by expert referees to the professional and scientific standards expected of a proceedings journal published by IOP Publishing.

Dosimetric Correction for a 4D-Computed Tomography Dataset using the Free-Form Deformation Algorithm

Daniel Markel, Hamideh Alasti and James C L Chow

Department of Radiation Oncology, University of Toronto and Radiation Medicine Program, Princess Margaret Hospital, Toronto, Ontario M5G 2M9, Canada.

james.chow@rmp.uhn.on.ca

Abstract. A Free-Form Deformable (FFD) image registration algorithm in conjunction with 4D Computed Tomography (CT) images was implemented within a graphical user interface, FFD4D, for dosimetric calculations. The algorithm was developed using the cubic-B-spline method with smoothness corrections and registration point assistance to mark fudicials. Validation of the algorithm was performed with manually measured geometric differences using a QUASAR Respiratory Motion Phantom. In this work, we used the FFD algorithm to demonstrate dosimetric corrections amongst 10 breathing phases of a lung cancer patient using the 4D-CT image datasets. Different methods to enhance the image processing speed for high-performance computing were also discussed.

1. Introduction

High-performance computing for medical imaging becomes popular as recent advances in medical imaging have enabled medical staff to create images of anatomy with unprecedented resolution and realism. The efficiency of high-performance computing in handling large image datasets makes the study of internal organ motion through image registration possible. In radiotherapy, interfraction and intrafraction organ motion can complicate treatment planning and execution. Localized lung motion in the order of 3 cm and prostate motion of up to 2.4 cm has been measured [1,2]. Breathing motion with amplitude of 1.5 cm has been shown to reduce tumour control probability (TCP) values by 3% [3]. Common practice when treating lung tumours is to combine the gross target volume (GTV) from breathing phases containing the displacement maxima into one planning target volume (PTV) to ensure complete coverage [4]. One drawback to this strategy is increased exposure to healthy tissue, which may cause higher toxicity and result in a higher risk of radiation pneumonitis [5]. An alternative strategy that has been investigated is the use of a margin which only accounts for set-up error. Predictably this results in a decreased risk to healthy tissue but at the expense of tumor coverage [6]. Precise and adaptive techniques appear to be the direction in which modern radiotherapy is headed. However, one of the shortcomings of most treatment planning systems such as Pinnacle[3] (Philips Medical Systems, Andover, MA) is the lack of a proper image registration tool for dosimetric calculations of plans involving non-rigid targets. There are generally few reliable and freely available image registration programs that go beyond rotating and translating the source image. In addition, image registration is time-consuming because it needs to process a group of large patient's image datasets varying with time. High-performance computing is therefore necessary in image registration with an acceptable computing time (within hour instead of day) for clinical treatment evaluation.

In the case of lung treatment without a proper method of accounting for all phases of breathing motion, dose calculations are performed on a single phase of the breathing cycle, which introduces errors of up to 10% in dose-volume histogram (DVH) calculations [7]. Three dimensional image registration has started to become increasingly available in medical imaging as a method of accounting for internal organ motion. Intrinsic image registration techniques generally fall into two categories, global and local. Global registrations involve the use of a single mathematical expression applied to the entire image. Such techniques include rotational, translational, affine, projective or conformal transformations. The parameters involved in global registrations are few and optimization of such parameters can be done quickly. Local registration techniques are used to account

Published under licence by IOP Publishing Ltd

for non-uniform differences between images, which involve adopting different transformation parameters for each region of the source image.

Free-Form Deformation (FFD) registration methods involve the use of a control grid overlaid on the images. Figure 1 demonstrates the concept by deforming a cube, using a surrounding control lattice. Each control grid vertex is considered a parameter and optimized according to an error function. Performing more complex techniques such as the Finite Element Method (FEM) has been preferred for non-elastic deformation.[7,8] However, one of the drawbacks of using the FEM is the difficulty in obtaining physical characteristics of the organs being deformed. The FFD algorithm requires no physical data other than the images. The software used in this study was created using a cubic-B-spline algorithm combined with smoothness, feature point, folding, maximum voxel displacement and volume change correction functions [9-11].

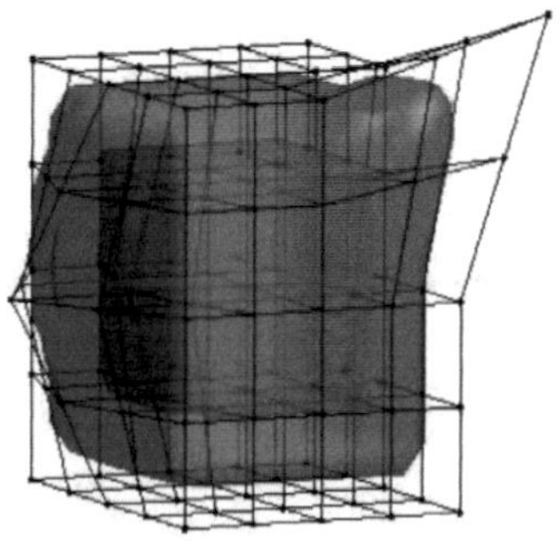

Figure 1. A hollow cube deformed using the FFD algorithm. Control grid is overlaid in black.

2. Methods

2.1. Theory

The in-house software developed for dose tracking relies on a cubic-B-spline FFD algorithm taken from Huang et al [11]. The algorithm makes use of a control grid overlaid on top of the image with control grid parameters (vertices) that are moved in order to warp the space, deforming the original data. The B-spline interpolation technique is defined by the following formula

$$L(\Theta;\mathbf{x}) = \sum_{q=0}^{3}\sum_{l=0}^{3}\sum_{r=0}^{3} B_q(u)B_l(v)B_r(w)(P^0_{i+q,j+l,k+r} + \delta P_{i+q,j+l,k+r}) \qquad (1).$$

L defines the new coordinates of the pixel $\mathbf{x}$ representing the coordinates of the initial position of the pixel in question, and P^0 (unchanged control grid points) and δP (change of the control grid points) define the final control grid parameters, Θ. This formula is used to determine the three error functions used in the optimization routine. These include the data driven error as well as a smoothness correction and feature point error. They are as follows.

$$E_{data}(\Theta) = \iiint_\Omega (\Phi_T(\mathbf{x}) - \Phi_S(L(\Theta;\mathbf{x})))^2 d\mathbf{x} \qquad (2)$$

$$E_{smooth}(\Theta) = \iiint_\Omega \left(\left\| \frac{\partial}{\partial x}\delta L(\Theta;\mathbf{x}) \right\|^2 + \left\| \frac{\partial}{\partial y}\delta L(\Theta;\mathbf{x}) \right\|^2 + \left\| \frac{\partial}{\partial z}\delta L(\Theta;\mathbf{x}) \right\|^2 \right) d\mathbf{x} \qquad (3)$$

$$E_{feature}(\Theta) = \sum_i (L(\Theta;\mathbf{x}_{si}) - \mathbf{x}_{di})^2; i \in [1, n_c] \qquad (4)$$

The three functions are combined using weighting factors α and β in Eq. (5).

$$E(\Theta) = E_{data}(\Theta) + \alpha E_{smooth}(\Theta) + \beta E_{feature}(\Theta) \qquad (5)$$

The optimization routine uses a gradient-based method according to Eq. (5).

2.2. Algorithm optimization and GUI

The goal of the deformation algorithm is to calculate the control grid parameters that will reproduce a source image to match a target image. The source and target images are assumed to be globally registered with coinciding resolutions at different phases of motion. Once calculated, the control grid is interpolated to the resolution of the source and target images to perform one-to-one voxel tracking and dose recalculation.

The image optimization process is summarized in Fig. 2. Initially, the selected anatomical structures on the DICOM images are contoured to outline the surfaces of target organs. This can be done automatically using edge detection functions in MATLAB or input through contours exported from Pinnacle[3]. It is these binary images that are run through the algorithm and optimized. Calculated control grids can be used afterwards to deform the original DICOM images. The first step in the optimization routine is to calculate the error function (Eq. (5)), which is based on three terms; least squares difference (calculated using the distance transforms of the binary contour images), a smoothness function (based on the gradient of the control grid) and error based on mismatching of feature points, which are manually designated by the user. The gradients of the control grid parameters are calculated analytically from the derivative of the error function and are subsequently used in a gradient-based optimization method to determine the correct control grid parameters that minimize the error function. After calculating an initial control grid, it is used to deform the original binary image of the source contour and calculate the next smoothness function. From the deformed source image and new position of the corresponding feature points, an adjustment to the control grid can be calculated and tested with varying step size, until it improves the value of the error function. Once this is done the process is repeated. After a preset number of iterations, the calculated control grid is outputted as the final deformation parameters. For registration of finer details of the images, the calculated control grid is interpolated to a higher resolution mesh and the optimization routine is repeated. Routines that are designed for creating physically realistic deformations and prevention of mesh folding were also implemented [11].

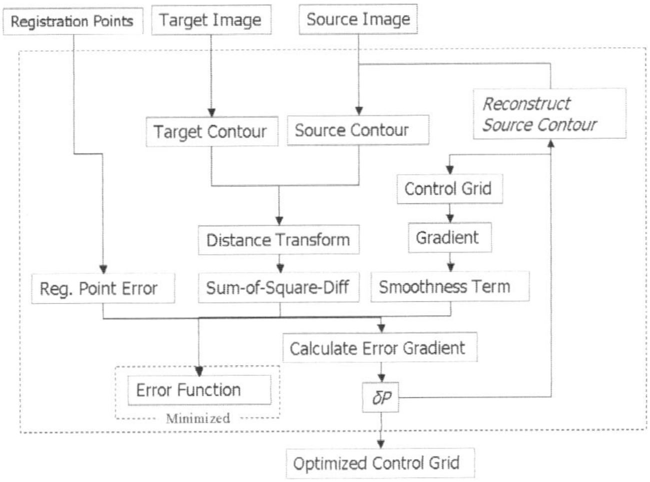

Figure 2. Flow chart of the cubic-B-spline FFD optimization routine used to register image sets. The routine was taken from Huang et al [11].

The main GUI of the FFD4D algorithm is designed using GUIDE on MATLAB as shown in Fig. 3. The big middle, top right-hand and bottom right-hand windows displayed the transverse, sagittal and coronal views of the lung patient's CT images. The vertical scrollbar in the left-hand side was used to select the corresponding breathing phase (0 – 90%) for image processing. Image dataset of a breathing phase can be registered through deformation based on selected feature points on the CT images in the GUI (blue, green and purple buttons). Patient's DICOM-RT and RTOG images are loaded into the software. Contours created in Pinnacle[3] and exported inside RTOG files are viewed and included in the patient data for use with the registration routine.

Control grid corrections were implemented from Huang et al. in addition to the basic registration algorithm [11]. First, a mesh-folding correction was applied. Mesh folding is a common problem that prevents proper warping and reconstruction of an image and causes voxels of a CT image to disappear, which is physically incorrect when applied as a model for tissue deformation [12, 13]. A check is performed on each vertex of the control grid lattice to ensure that it falls within a boundary created by the nearest surrounding control points so that further folding is not created when interpolating the control grid to a finer mesh. A routine was designed to check that all control points fall within the octahedron designated by the surrounding points. Points that fell outside this space were moved to the centre of the octahedron as opposed to their original position without motion as outlined in the literature, to preserve the general shape of the deformation [11].

Another correction is made to ensure that motion does not exceed the maximum voxel displacement [12]. The maximum voxel displacement was set to 4 cm per phase for all voxels, which is an acceptable upper limit for lung motion [1]. To implement it for the deformation vector of each vertex $\delta\vec{P}_{m,n,o}$ in the control grid lattice is multiplied by the factor,

$$\alpha_{MVD} = -\exp\left(-0.1\left(\frac{\left|\delta\vec{P}_{m,n,o}\right|}{MVD(m,n,o)}\right)^{6}\right) \quad (6)$$

This ensures that the deformation vector's amplitude and subsequently voxel motion never exceeds 4 cm even if it avoids causing mesh folding. This forces a more anatomically correct deformation.

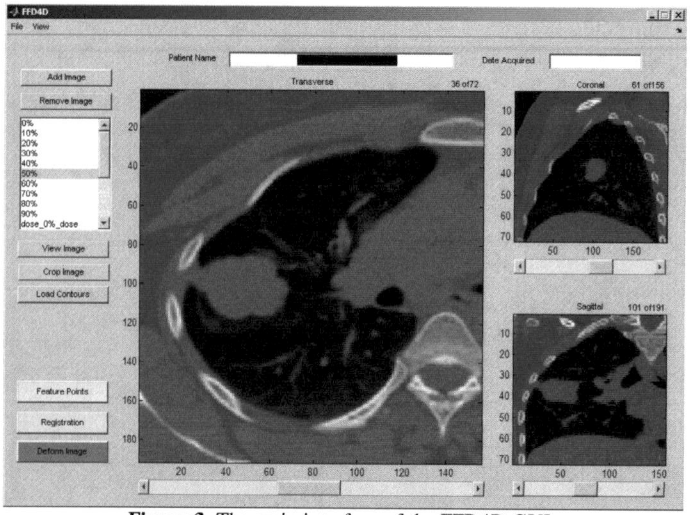

Figure 3. The main interface of the FFD4D GUI.

2.3. FFD Validation and 4D-CT image deformation

For validation purposes, a QUASAR respiratory motion phantom was chosen, as the changes due to motion are detectable and geometric features are easily identifiable. The phantom was moving 4 cm in the super-inferior directions to simulate the diaphragm motion of a patient due to free breathing motion [14]. The phantom has a cylindrical cedar insert (mimicking lung density), held within an acrylic oval body. One tissue-equivalent cubic ($3 \times 3 \times 3$ cm^3), and one spherical insert (1 cm diameter) are embedded in the cylindrical lung cedar insert. The "oscillation" mode of the phantom was used and a patient's breathing profile was programmed into the motor to move the inserts according to the patient's breathing pattern. The phantom was imaged in 10 phases using a GE light speed plus CT simulator (16 slices at 60 kVp), capable of 4D-CT imaging [15]. The 4D-CT images were binned into 10 phases along the breathing cycle, and each phase was registered to 30%. The inserts were then contoured using the Pinnacle[3] treatment planning system, with a feature point designated at the center of the spherical insert. Geometric features of the inserts were compared using the GUI's feature point sub-menu due to its ability to measure points in space from all transverse, sagittal and coronal planes. Images of the QUASAR respiratory phantom at all 10 phases were interpolated to $100 \times 100 \times 80$ pixels prior to registration. The control grid lattice was first optimized at a resolution of $22 \times 22 \times 22$ pixels at 5 iterations, then at a finer resolution of $42 \times 42 \times 42$ pixels for another 5 iterations. All 10 phases were chosen for comparison using the corners of the cubic insert and the center of the spherical insert as points of comparison for a total of 81 points (9 points $\times$ 9 comparisons).

For a 3D standard lung protocol at the Princess Margaret Hospital, the GTV at 0% and 30% phases are fused and a margin of 6 mm is added to define the PTV [4, 16]. This protocol does not incorporate image deformations in the dosimetry calculations. The in-house developed FFD algorithm was used to compare the dosimetric data in our hospital versus all 10 phases of the breathing cycle after having been registered to the 30% phase. The tumor and lung were delineated at all 10 phases using Pinnacle[3] treatment planning system and then exported to the FFD4D program as shown in Fig. 3.

3. Results and discussion

Figure 4 demonstrates the accuracy of the registration software. Figure 4(a) shows the cubic insert at the 10% breathing phase, and Fig. 4(b) shows the same insert at 30% phase after a 4 cm motion in the superior direction. Figure 4(c) shows the image of the insert from the 10% phase after registration and deformation to the 30% phase. Table 1 summarizes the results of the 81-points of comparison taken from 10 phases in the respiratory motion phantom.

Dosimetry was calculated from a 6-beam intensity modulated radiotherapy (IMRT) plan treating a tumour in the right mid-section of the lung as shown in Fig. 5(a) [17]. The FFD4D software was used to determine the differences in a patient's DVHs when deformation and motion are considered during 10 phases of breathing. The IMRT plan based on this standard protocol was copied to the other 9 phases image sets, and dosimetry was recalculated using the CT data of those phases [4, 16]. The dose data was then reconstructed using the transformation parameters calculated through registration of the contours of selected anatomical structures for each phase as shown in Fig. 5(b). The registered 10 sets of dose data were voxel-tracked and DVHs were calculated using the contours delineated in the 30% phase plan. DVH of these PTVs were compared with the voxel tracked one as shown in Fig. 6(a). The variability due to organ motion is clearly visible from phase to phase. Figure 6(b) compares a voxel-tracked DVH to that of the standard protocol demonstrating the reduction in the dose coverage of the PTV. Figure 6(b) shows a peak difference of 10 cm^3 at 4400 cGy between the DVH based on the PTV as per the standard protocol and the DVH taken when motion of all 10 phases is accounted for [17].

It is generally difficult to comparatively validate registration software, as there are many methods of doing so and no gold standard [18]. The accuracy of FFD registrations has also been found to vary for different regions of the body, and is patient specific [18]. A trained physician or therapist is also necessary to properly identify the coordinates of anatomical features. For these reasons a phantom was chosen for validation. From geometric comparisons of the registered image sets, a total average error of 3.2 mm was found with a range of 1.7 – 4.3 mm (Table 1). This is comparable and on par with other registration programs. Other FFD algorithms have reported error ranges of 0 - 6 mm [18-22]. FEM software has reported error averages of 2.4 - 4.4 mm [7, 8]. Table 1 shows that errors in the x and y (left-right and anterior-posterior) directions are minimal and only exist because of the effect of image blur on the auto contouring filter used in Pinnacle[3] to draw the insert contours. Ideally they should not exist as the phantom motion was designated solely in the superior/inferior axis. The motion of the cedar inputs was also a purely rigid transformation, which is possible yet difficult for large amplitudes under FFD. Unfortunately no phantom yet exists that includes deformable inserts for this study, which is why the QUASAR phantom was chosen. Error in the z (superior-interior) direction tends to increases with difference in starting positions, as larger deformations are necessary to register the two contours. At greater distances, registration points improve the deformation significantly.

The registration process required approximately 160 minutes per phase using a PC with a Pentium 4, 3.20 GHz processor, 1 GB RAM running MS-WINDOWS 2000 32-bit and MATLAB 2008a. When the process was run on the General Purpose Cluster at SciNet (University of Toronto), all registrations of 9 phases could be run in parallel. This greatly improved the computing time. However, to further increase the computing efficiency apart from parallel processing, the FFD algorithm was investigated to find out if high performance can be achieved by reduction of data space and reduction in solution search space. Control grids were optimized at two grid sizes of 22×22×22 and 42×42×42 for 5 iterations each. Application of deformations required approximately 1-5 minutes depending on the resolution of the deformed dataset. Calculations using FFD algorithms are well known to be computationally intensive. Changing the registration settings to reduce the iterations in the optimization routine, or using substantially less control grid parameters (e.g. using 12×12×12) can significantly reduce the calculation time to as little as 30 minutes in the case of FFD4D. However the detail of the registration is often sacrificed in this case. Other implementations of the cubic-B-spline algorithm have reported calculation times of around 70-100 minutes [20, 21]. One method of reducing calculation times is to use multi-resolution algorithms [21-23] which vary the resolution of the deformation parameters based on the magnitude of the displacement. This has proven to reduce calculation times significantly to as much as half an hour per phase depending on the parameters [20]. Using alternative error criteria and tailored hardware architecture have shown to reduce calculation times such that FFDs are appropriate for clinical implementation, with some registration times being as fast as 3 minutes [21, 24].

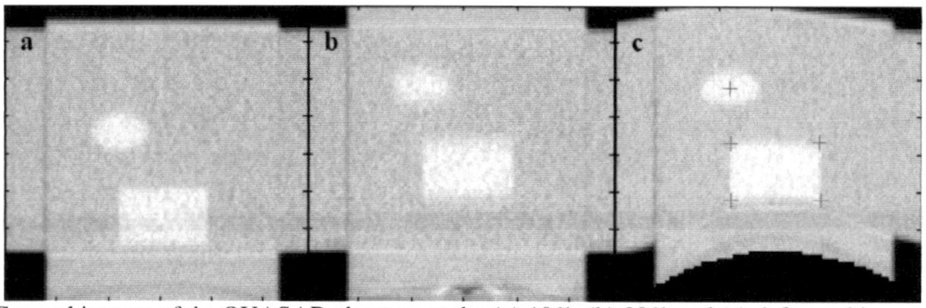

Figure 4. Coronal images of the QUASAR phantom at the (a) 10%, (b) 30% and (c) deformed image of (a) after registration. Red cross hairs designate points of comparison.

Table 1. Error from 9 phases after registration to the 30% phase.

Phase	Δx (L-R) Average (mm)	Δy (A-P) Average (mm)	Δz (S-I) Average (mm)	Total Distance Average (mm)
0%	0.230	0.034	1.694	1.737
10%	0.656	0.344	2.252	2.443
20%	0.064	0.000	4.180	4.185
40%	0.000	0.000	2.717	2.717
50%	0.138	0.000	2.194	2.236
60%	0.061	0.000	4.082	4.085
70%	0.069	0.000	4.284	4.287
80%	0.206	0.000	3.344	3.376
90%	0.122	0.000	3.993	4.001
			Final Average	**3.229**

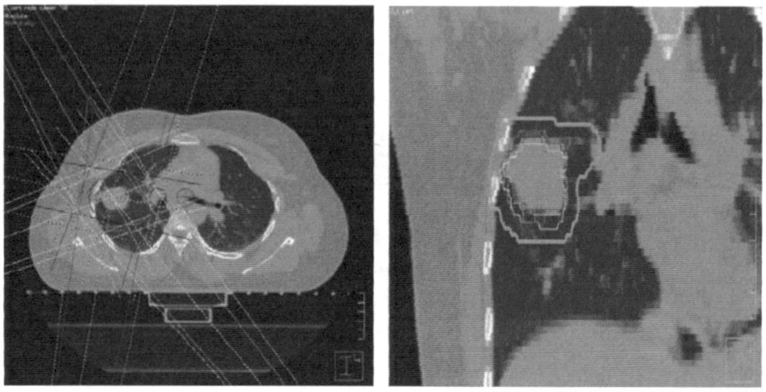

(a) (b)

Figure 5. (a) A transverse view of the patient displaying the 6-beam IMRT plan. (b) A coronal view of the patient showing the GTV contours of all 10 phases superimposed onto the 30% CT data as well as the PTV in orange colour.

(a)

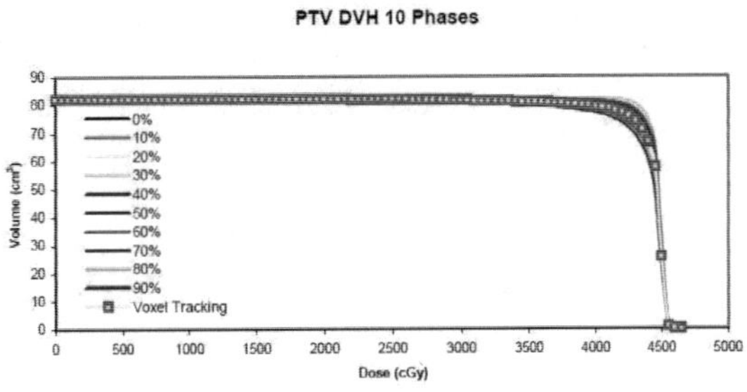

(b)

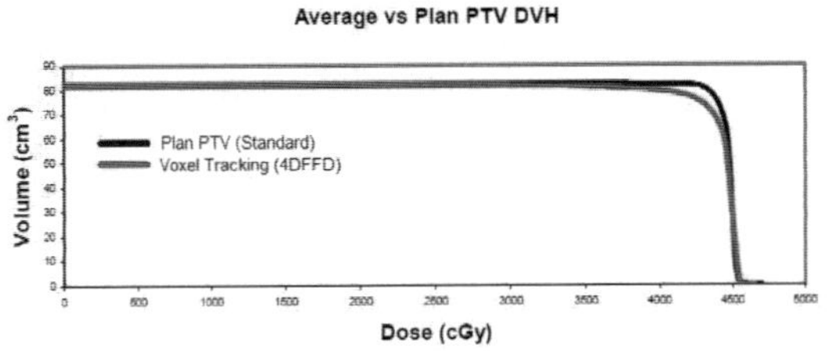

Figure 6. (a) DVHs of the PTV using dose data from all 10 phases registered to the 30% phase and (b) comparison of the average DVH from (a) for the PTV, compared to that taken solely from 30% plan.

4. Conclusions

A FFD algorithm was developed that includes smoothness correction, and feature point error. Maximum voxel and volume displacement prevention corrections along with checks for mesh folding have also been integrated into the software. This code was encapsulated into a GUI called FFD4D using MATLAB. Validation of the FFD algorithm was performed using a QUASAR Respiratory Motion Phantom due to the ease at which matching features at different phases in the motion cycle could be identified. It was found that the registrations performed on 10 phases of phantom yielded an average error of 3.2 mm, with a range of 1.7 – 4.3 mm. A demonstration of applying the software to patient dosimetry was also performed. It is concluded that performing the image registration of different breathing phases in parallel using high-performance computing cluster can enhance the image processing speed, and the FFD algorithm can produce a realistic dose-volume relationship of a lung patient due to the breathing motion in radiotherapy.

References

[1] Berbeco R I, Nishioka S, Shirato H, Chen G T, Jiang S B. 2005 Residual motion of lung tumours in gated radiotherapy with external respiratory surrogates, Phys. Med. Biol. **21**, 3655-67.

[2] Kitamura K, Shirato H, Seppenwoolde Y, Onimaru R, Oda M, Fujita K, Shimizu S, Shinohara N, Harabayashi T, Miyasaka K. 2002 Three-dimensional intrafractional motion of prostate measured during real-time tumour-tracking radiotherapy in supine and prone treatment positions, Int. J. Radiat. Oncol. Biol. Phys. **53**, 1117-1123.

[3] Engelsman M, Damen M F E, De Jaeger K, van Ingen K M, Mijnheer B J. 2001 The effect of breathing and set-up errors on the cumulative dose to a lung tumor, Radiother. Oncol. **60**, 95-105.

[4] Alasti H, Cho Y B, Vandermeer A D, Abbas A, Norrlinger B, Shubbar S and Bezjak A. 2006 A novel four-dimensional radiotherapy method for lung cancer: imaging, treatment planning and delivery, Phys. Med. Biol. **51**, 3251-3267.

[5] Rodrigues G, Lock M, D'Souza D, Yu E and Van Dyk J. 2004 Prediction of radiation pneumonitis by dose–volume histogram parameters in lung cancer—a systematic review, Radiother. Oncol. **71**, 127–38.

[6] Colgan R, McClelland J, McQuaid D, Evans P M, Hawkes D, Brock J, Landau D and Webb S. 2008 Planning radiotherapy using 4D CT data and a motion model, Phys. Med. Biol. **53**, 5815-5830.

[7] Kung J H, Zygmanski P, Choi N, Chen G T Y. 2003 A method of calculating a lung clinical target volume DVH for IMRT with intrafractional motion, Med. Phys. **30**, 1103-1109.

[8] Brock K K, Sharpe M B, Dawson L A, Kim S M, and Jaffray D A. 2005 Accuracy of finite element model-based multi-organ deformable image registration, Med. Phys. **32**, 1647-1659.

[9] Schnabel J A, Tanner C, Castellano A D, Degenhard A, Hayes C, Leach M O, Hose D R, Hill D L G, Hawkes D J. 2001 Validation of non-rigid registration of contrast-enhanced MR mammography using finite element methods, Lecture Notes in Computer Science, **2082**, 344-357.

[10] Huang X, Paragios N, Metaxas D. 2008 Shape registration in implicit spaces using information theory and free form deformations, IEEE Transactions on Pattern Analysis and Machine Intelligence. **28**, 1303-1318.

[11] Huang X, Zhang S, Wang Y, Metaxas D, and Samaras D. 2004 A hierarchical framework for high resolution facial expression tracking, Proc. Third IEEE Workshop Articulated and Non-rigid Motion, in conjunction with CVPR '04.

[12] Castro-Pareja C R and Shekhar R. 2004 Physically correct mesh manipulation in multi-level free-form deformation based nonrigid registration, Proc. of the 26th Annual Inter. Conf. ff the IEEE EMBS.

[13] Castellano-Smith A D, Hartkens T, Schnabel J, Hose D R, Liu H, Hall W A, Truwit C L, Hawkes D J and Hill D L G. 2008 Constructing patient specific models for correcting intraoperative brain deformation, Lecture Notes in Computer Science, **2208**, 1091-1098.

[14] Publicover J, Vandermeer A, Norrlinger B, Alasti H. 2008 Feasibility of using a programmable respiratory motion phantom for QA and assessment of dosimetric implications of breathing motion during radiation therapy, Med. Phys. **35**, 2795.

[15] Castellano-Smith A D et al. 2002 A registration based mesh construction technique for finite element models of brains, Proc. SPIE **4684,** 538-549.

[16] ICRU "Prescribing, recording and reporting photon beam therapy (supplement to ICRU report 50) ICRU report 62" (Bethseda, MD: International Commission on Radiation Units and Measurements) (1999).

[17] Alasti H, Chow J, Markel D, Lochovsky C. 2008 Evaluation of a 4DCT-based IMRT plan for lung using the free-form image deformation dose tracking method, Int. J. Radiat. Oncol. Biol. Phys. **72,** S614-5.

[18] Holton K S, Taneja U, Robb A R. 1995 Quantitative validation of 3D image registration techniques," Proc. SPIE, **2434**, 504.

[19] Mattes D, Haynor R D, Vesselle H, Lewellen K T, and Eubank W. 2003 PET-CT image registration in the chest using free-form deformations, IEEE Trans. Med. Im. **22**, 120-128.

[20] Declerck J, Denney T S, Ozturk C, O'Dell W and McVeigh E R. 2000 Left ventricular motion reconstruction from planar tagged MR images: a comparison, Phys. Med. Biol. **45**, 1611-32.

[21] Dawant B M, Hartmann S L, Thirion J P, Maes F, Vandermeulen D and Demaerel P. 1999 Automatic 3-D segmentation of internal structures of the head in MR images using a combination of similarity and free-form transformations: part 1, methodology and validation on normal subjects, IEEE Transactions of Medical Imaging **18**, 917-926.

[22] Castadot P, Lee A J, Parraga A, Geets X, Macq B, Grégoire V. 2008 Comparison of 12 deformable registration strategies in adaptive radiation therapy for the treatment of head and neck tumors, Radiother Oncol. **89**, 1-12.

[23] Tanner C, Schnabel J A, Hill D L, Hawkes D J, Degenhard A, Leach M O, Hose D R, Hall-Craggs M A, Usiskin S I. 2007 Quantitative evaluation of free-form deformation registration for dynamic contrast-enhanced MR mammography, Med. Phys. **34**, 1221-33.

[24] Castro-Pareja C R, Jagadesh J M, Shekhar R. 2003 FAIR: A hardware architecture for real-time 3-D image registration," IEEE Transaction on Information Technology in Biomedicine, vol. **7**, no. 4, pp. 426-434.

Challenges in analysing and visualizing large-scale molecular dynamics simulations: domain and defect formation in lung surfactant monolayers

E Mendez-Villuendas, S Baoukina, D P Tieleman

Department of Biological Sciences and Institute of Biocomplexity and Informatics, University of Calgary

E-mail: tieleman@ucalgary.ca

Abstract. Molecular dynamics simulations have rapidly grown in size and complexity, as computers have become more powerful and molecular dynamics software more efficient. Using coarse-grained models like MARTINI system sizes of the order of 50 nm x 50 nm x 50 nm can be simulated on commodity clusters on microsecond time scales. For simulations of biological membranes and monolayers mimicking lung surfactant this enables large-scale transformation and complex mixtures of lipids and proteins. Here we use a simulation of a monolayer with three phospholipid components, cholesterol, lung surfactant proteins, water, and ions on a ten microsecond time scale to illustrate some current challenges in analysis. In the simulation, phase separation occurs followed by formation of a bilayer fold in which lipids and lung surfactant protein form a highly curved structure in the aqueous phase. We use Voronoi analysis to obtain detailed physical properties of the different components and phases, and calculate local mean and Gaussian curvatures of the bilayer fold.

1. Introduction

Computer simulations of models of biological membranes and monolayers have a long history [1, 2]. They have evolved from simulations of a few dozen simplified lipids in a clearly defined lamellar geometry to very complex systems including a variety of lipids and proteins in geometries that can be quite complex [2, 3]. This increase in size and complexity is driven by increases in computer power and more efficient simulation software, but received a major boost with the advent of coarse-grained models that are up to several orders of magnitude faster than atomistic simulations. The popular MARTINI model used in this paper allows simulations of a scope 2-3 orders of magnitude larger than atomistic simulations for the same computational cost, at the obvious expense of a loss of detail [4, 5]. In practice, simulations with a size of the order of 50 x 50 x 50 nm can now be run relatively easily on a few hundred CPU cores for simulation times of the order of 10 microseconds. At these time and length scale, bilayer and monolayers can show major changes in organization. Combined with increasingly complex compositions this poses challenges for analysis, as properties of interest vary throughout the inhomogeneous system and cannot be calculated by simple averaging. Instead, methods to identify regions of interest first, followed by identification of the molecules in that region, are required to do further analyses.

In addition, there is a growing disparity between the size of a typical system and the smallest interactions of interest. In a typical bilayer simulation of 15 years ago, the system contained 128 lipids plus water, in a box that was ca. 6 x 6 nm laterally, while an individual lipid has a surface area in the membrane of ca. 0.6 nm^2. To zoom from a view of the whole system to a view of individual lipids requires about a 10-50-fold 'magnification', or, in an alternative point of view, zooming from a view that shows 15,000 atoms to a view that shows 100. Current systems can be of the order of 100 x 100 x 100 nm, with millions of atoms, with a much larger difference in scale between the whole system and molecular details, while both are still of interest.

In this report we examine a simulation of a scope and complexity we think will become typical in the next few years, even if it is currently at the edge of what is feasible. In this simulation, as in several previously published papers, we are inspired by questions related to the function of lung surfactant.

Lung surfactant is a monomolecular film that covers the interface of the alveoli and air. The alveoli are small structures in the lung where the actual gas exchange takes place. During the breathing cycle the area of this interface changes substantially. Upon inhalation the alveoli expand, increasing the area of the interface, while during exhalation the opposite occurs. At most interfaces, for instance the air/water interface, changing the area requires a significant amount of work, expressed quantitatively by the surface tension at the interface. The main role of lung surfactant is to reduce the surface tension to very low values, so that breathing does not require significant work [6]. Lung surfactant has several components but consists primarily of the zwitter-ionic lipid dipalmitoyl-phosphatidyl choline (DPPC), and contains anionic lipids, unsaturated lipids and cholesterol. In addition, there are a number of proteins associated with lung surfactant, two of which are thought to play a critical role in its primary function: surfactant-associated proteins B and C (SP-B, SP-C) [7].

In previous simulation studies of lung surfactant, we focused on a number of model systems. We studied monolayers formed by DPPC [8], a more realistic lipid mixture of DPPC, palmitoyl-oleoyl-phospatidylglycerol (POPG), and cholesterol and surfactant proteins to test two different models of lung surfactant reservoirs [9], the general mechanism of single-phase monolayer breakdown [10]; stability conditions for monolayers [11], and the ability of SP-B to facilitate fusion and lipid flow between a monolayer and bilayers/vesicles [12, 13]. However, these simulations, by design, ignored the possibility that lung surfactant monolayers may contain coexisting liquid-disordered and liquid-ordered phases. We also did not analyze in any detail the distribution of different lipids near irregular features such as the connection points of vesicles to a monolayer or fusion intermediates. Here we describe methods to analyze a mixture mimicking lung surfactant under conditions of surface tension and temperature where there is clear evidence of domain formation, protein sorting, and eventually monolayer breakdown or defect formation.

2. Methods

The simulation here uses the same methods previously described [13]. We refer to that paper for details of the MARTINI force field and simulation parameters. The system setup consisted of a water slab in vacuum with two symmetric monolayers at the two water/vacuum interfaces. The monolayers consist of dipalmitoyl-phosphatidylcholine (DPPC), dioleoyl-phosphatidylcholine (DOPC), palmitoyloleoyl-phosphatidylglycerol (POPG), and cholesterol in ratio 2:1:1:1, respectively. Two hydrophobic surfactant proteins SP-B and SP-C were added to the monolayers. Each monolayer contains 2304 lipids, and 6 SP-B and 8 SP-C proteins. The water slab contains 124,000 water particles, of which 398 were substituted for Ca^{2+} ions to neutralize the unbalanced charge resulting from anionic POPG lipids and net positively charged proteins.

The lipids were randomly mixed in the starting configuration; SP-B proteins were placed in water and SP-C proteins in vacuum in the vicinity of each monolayer. Initially the monolayers were equilibrated for 1 μs at a surface tension of 30 mN/m and a temperature of 290 K, at which the monolayers remained roughly homogeneous and stable at the interface. After equilibration, the surface tension was reduced to 10 mN/m, which led to lipid segregation and monolayer collapse from the interface via formation of a bilayer fold induced by SP-B proteins.

To analyze the frames from the trajectory, we developed a Matlab-based (v.R2011b) analysis program to perform quantitative analysis of complex mixtures based on Voronoi tessellation [Mendez-Villuendas et al, in preparation]. In the figures this is based on 2D tessellation using specific lipid particles as centers of Voronoi polygons: the R1 site in cholesterol (near the OH headgroup), and the C1A and C1B sites in the phospholipids, the first beads of the hydrocarbon chain (two Voronoi cells per lipid).

During the simulation, two distinct phases appear. We assigned each Voronoi cell to a specific phase using a cluster analysis based on a connectivity matrix with a cutoff parameter. The cutoff corresponded roughly to the square root of the average area of the Voronoi polygons in the more ordered phase. In addition, lipids belonging to a more ordered phase were required to have an average chain order parameter, Sz, larger than a selected value. Sz is a measure of the orientation of a lipid chain – a high value corresponds to a gel or liquid-ordered domain, while a lower value corresponds to a liquid expanded or liquid-disordered domain.

The monolayer curvature was characterized using a Matlab-based (v.R2011b) program (Mendez-Villuendas et al, in preparation). The phosphate groups of lipids (PO4 sites) were fitted to a surface using a binomial filter in 2D with a 6 nm characteristic length to remove the noise. This convoluted surface was then converted to an equally-spaced grid (0.6 nm spacing), for which partial derivatives were calculated to find the principal curvatures, c1 and c2, the mean curvature $H = ½ (c_1+c_2)$ and the Gaussian curvature $K = c_1c_2$

3. Results and Discussion

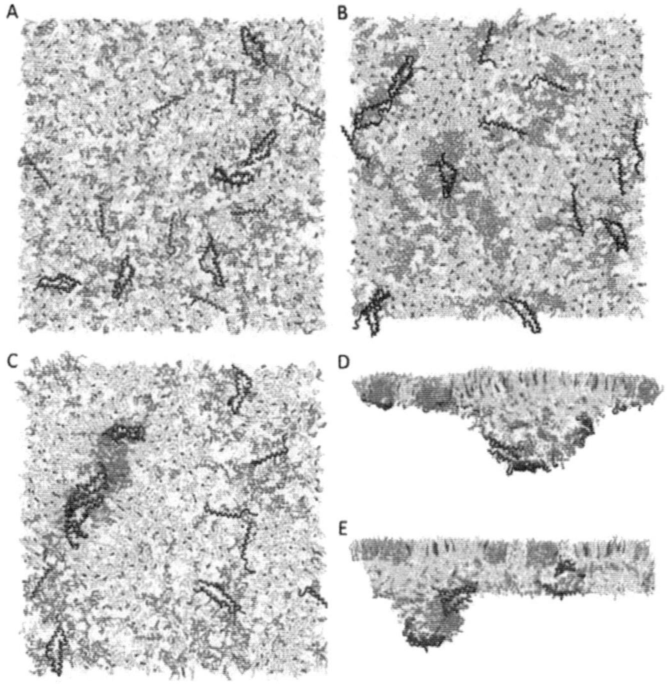

Figure 1. (A) Roughly homogeneous mixture after 1 microsecond at a surface tension of 30 mN/m and 290 K. (B) Lipid components segregate and SP-Bs form a dimer after 7 microseconds at 10 mN/m and 290 K. (C) SP-B dimer produces a bilayer fold connected to monolayer after 8.65 microseconds. (D-E) side views on the bilayer fold.

Color scheme: DPPC in green, DOPC in orange, POPG in yellow, cholesterol in purple, SP-B in red, SP-C in mauve. Lipid tails are shown as sticks, polar groups as beads, protein backbone as licorice, side chains not shown, water not shown.

Figure 1 shows a number of snapshots of the simulation. The snapshots 1A, B, C show a view from the water phase normal to the monolayer. At t=0, the system is in a metastable state, induced from a stable state by lowering the surface tension below the equilibrium. Figure 1B shows that after ca. 7 microseconds, there is clear evidence of phase separation and domain formation. The surfactant

Journal of Physics: Conference Series **385** (2012) 012002 doi:10.1088/1742-6596/385/1/012002

proteins preferentially interact with DOPC/POPG components forming a liquid-disordered (Ld) phase, while DPPC and cholesterol separate out into a liquid-ordered (Lo) domain with occasional inclusions of POPG, preferentially at the boundary. Surfactant proteins induce local perturbation of the monolayer.

The monolayer at the interface is still metastable, and a significant defect forms after 8 microseconds (Figure 1C), which can be clearly seen in the side views (along two different axes) of this monolayer in Figures 1D and 1E. The defect represents a bilayer fold of a disc-like shape, connected to a monolayer. The bilayer fold is induced by an SP-B dimer via local bending of the monolayer. We previously observed formation of a bilayer fold from a monolayer below the equilibrium surface tension induced by SP-Bs in the absence of phase separation [9]. Here it is clear even without numerical analysis that the fold nucleates in the Ld phase and becomes highly enriched in POPG/DOPC/protein, with little cholesterol or DPPC.

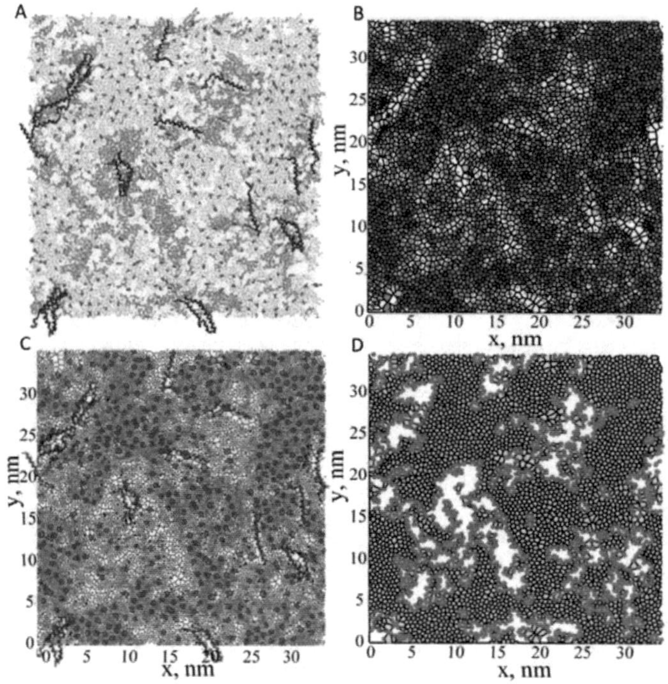

Figure 2. Voronoi tessellation of the monolayer in Figure 1B. Color scheme as in Figure 1; Voronoi polygons are colored according to lipid type. Proteins are not included in the tessellation. (A) Snapshot of the monolayer. (B) Voronoi lattice of the same structure. (C) Voronoi lattice combined with protein structures. (D) Domains based on identifying two phases (blue and white; red indicates the domain boundaries).

Figure 2 illustrates how the Voronoi analysis is helpful in analyzing this complex mixture. We considered (Figure 2A) a flat monolayer with Ld and Lo domains (at 7 μs) to characterize in-plane distribution of components. Figure 2B shows 2D Voronoi tessellation based on selected lipid sites only (see Methods). Once lipids have been assigned to a Voronoi cell, additional analysis becomes possible on subsets of cells that share common features. We can use the Voronoi polygons to calculate areas of individual lipids and to group lipids to domains of distinct phases. Figure 2C superimposes the protein structures on the Voronoi cells. It becomes apparent that lipids near the proteins have larger areas. We can readily identify lipids that are near proteins, determined by a distance criterion or other means, and analyze these as group.

Figure 3 shows variation of lipid properties as function of in-plane radial distance from the proteins based on Figure 2C. The areas per lipid (Figure 3A) increase noticeably in the protein vicinity due to partitioning of apolar protein residues in the monolayer. Interestingly, this increase is followed by a decrease in the area per DOPC lipid compared to its average area in the monolayer.

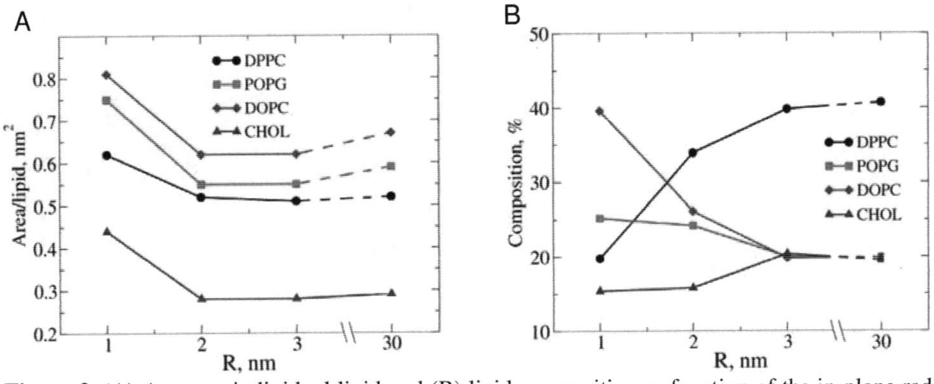

Figure 3. (A) Area per individual lipid and (B) lipid composition as function of the in-plane radial distance from surfactant proteins SP-B and SP-C. Lipid distribution around proteins was calculated by averaging over concentric shells of 1 nm increment. The points at 30 nm converge to the average values.

This is possibly due to an increase of its local concentration in the protein vicinity, which can be seen in the composition profile (Figure 3B). The distribution of lipid components also shows a moderate increase in POPG concentration near proteins, and a local depletion of saturated DPPC lipids. These differences in lipid concentration near proteins promote lipid segregation between the ordered and disordered phases. At large distances, the composition of the monolayer is reaches its average value, as expected.

A large number of other analyses are possible. We can trivially analyze averages over all DOPC lipids in the system, because we know the nature of all lipids in the simulation. However, DOPC lipids behave differently in different phases, and for more meaningful analyses we need to distinguish between DOPC lipids in Lo and Ld phases. From a detailed examination of the images it becomes obvious that lipid components do not fully segregate between the phases; the Ld phase, while highly enriched in DOPC, also contains POPG and a small fraction of DPPC and cholesterol. We can group lipids into Lo/Ld domains by building the connectivity matrix based on selected physical properties. These properties include lipid packing (which can be expressed by a distance cutoff or a number of neighbors), or chain order parameter. By analyzing the edges of the Voronoi polygons, we can also characterize the boundary between the phases/domains (Figure 2D).

Once domains have been identified, we can track the number of individual domains over time and characterize the kinetics of domain growth. We can analyze the properties that are essential for understanding the energetics of domains, such as the area of each phase and the length of the phase boundary, and dynamics, such as diffusion of lipids within each domain. We can also obtain other properties of each phase, including lipid composition, partial areas of each lipid component and order parameters. We can follow the response of all these properties to changes in e.g. temperature, composition, and surface tension.

We are extending the software to do a 3D tessellation, but for mostly flat geometries 2D is adequate. There is an interesting opportunity to combine the Voronoi analysis with other grid-based analyses. One key example is given by the Gaussian and mean curvatures that can be calculated from the principal curvatures. To calculate these curvatures we need to overlay the molecular system, which has a substantial roughness/noise at any given time, with a filter to smooth the surface, and then discretize it. From this smoothed surface on a mesh grid we can numerically calculate the partial derivatives that define the local curvature.

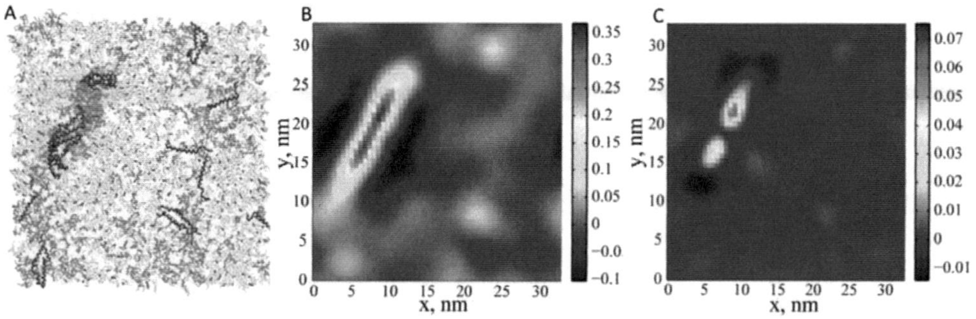

Figure 4. (A) Monolayer structure with bilayer fold (same as in figures 1CDE); (B) Mean and (C) Gaussian curvatures of the monolayer in (A).

Figure 4A again shows a snapshot of the monolayer system described in more detail in Figure 1. Figure 4B and C show the calculated mean and Gaussian curvature for the bilayer fold connected to the monolayer (at 8.65 μs), corresponding to Figure 1C-E, using a filter with a 6 nm resolution. The resolution of the filter is limited by the validity of the concept of curvature as a physically meaningful quantity for membranes on the low end and by the desired resolution of the analysis at the high end. Specifically, if the filter is too small curvature can be calculated for individual lipid molecules but this has no useful physical meaning, while if the filter is very large important features on smaller scales will be ignored. Thus there is significant room for further development as well as combining this curvature grid with the Voronoi grid.

4. Outlook

We have illustrated some current challenges in analyzing state of the art simulations of biological systems, representing complex lipid-protein 3D assemblies. The Voronoi analyses allow us to numerically distinguish different environments for the same molecules, so that properties can be calculated for a specific type of environment only, or the interfacial length between two phases, quantities that could be used to study the factors explaining the energy gain in one molecular mechanism over another. For example, proteins may play a role in reducing the free energy of formation of a new nucleation site via an energetic preference for a particular lipid, leading to the formation of a new phase. Further, models describing domain energetics may be linked to mesoscopic theories that have the overarching goal of explaining phenomena at the biological scale. Curvature plays a key role in this class of theories; the ability to link molecular simulation results through curvature to mesoscopic Helfrich-type models will open a new realm of applicability of molecular simulation and strengthen both large-scale lipid simulations and the use of continuum models.

Acknowledgements
This work was supported by the Canadian Institutes of Health Research (EMV) and the Natural Sciences and Engineering Research Council of Canada (SB). DPT is an Alberta Innovates Health Solutions Scientist and Alberta Innovates Technology Futures Strategic Chair in (Bio)Molecular Simulation. Simulations were carried out on WestGrid/Compute Canada facilities. This research was supported in part by the Project of Knowledge Innovation Program (PKIP) of Chinese Academy of Sciences, Grant No. KJCX2.YW.W10

Supplementary information: An animation of the simulation described here is available with the online version (m4222sp-10.290.avi). This movie shows a ~ 9 microsecond trajectory at a surface tension of 10 mN/m and a temperature of 290 K starting from the monolayer equilibrated at a 30 mN/m as described in the Methods. The color scheme as in Figure 1.

References

[1] Tieleman D P, Marrink S J and Berendsen H J C 1997 A computer perspective of membranes: Molecular dynamics studies of lipid bilayer systems *Biochim. Biophys. Acta* **1331** pp 235-270

[2] Marrink S J, de Vries A H, and Tieleman D P 2009 Lipids on the move: Simulations of membrane pores, domains, stalks and curves. *Biochim. Biophys. Acta Biomembranes* **1788** pp 149-168

[3] Marrink S J and Tieleman D P 2002 Molecular dynamics simulation of spontaneous membrane fusion during a cubic-hexagonal phase transition *Biophys. J.* **83** pp 2386-2392

[4] Marrink S J *et al* 2007 The MARTINI force field: coarse grained model for biomolecular simulations *J. Phys. Chem. B* **111** pp 7812-7824

[5] Monticelli L *et al* 2008 The MARTINI coarse-grained force field: extension to proteins *J. Chem. Theo. Comp.* **4** pp 819-834

[6] Rugonyi S, Biswas S C, and Hall S B 2008 The biophysical function of pulmonary surfactant *Respiratory Physiology & Neurobiology* **163** pp 244-255

[7] Zuo Y Y *et al* 2008 Current perspectives in pulmonary surfactant - Inhibition, enhancement and evaluation *Biochim. Biophys. Acta Biomembranes* **1778** pp 1947-1977

[8] Baoukina S *et al* 2007 Pressure-area isotherm of a lipid monolayer from molecular dynamics simulations *Langmuir* **23** pp 12617-12623

[9] Baoukina S *et al* 2007 The molecular mechanism of monolayer-bilayer transformations of lung surfactant from molecular dynamics simulations *Biophys. J.* **93** pp 3775-3782

[10] Baoukina S *et al* 2008 The molecular mechanism of lipid monolayer collapse *Proc. Natl. Acad. Sci. USA* **105** pp 10803-10808

[11] Baoukina S, Marrink S J, and Tieleman D P 2010 Lateral pressure profiles in lipid monolayers *Faraday Discussions* **144** pp 393-409

[12] Baoukina S and Tieleman D P 2010 Direct Simulation of Protein-Mediated Vesicle Fusion: Lung Surfactant Protein B *Biophys. J.* **99** pp 2134-2142

[13] Baoukina S and Tieleman D P 2011 Lung surfactant protein SP-B promotes formation of bilayer reservoirs from monolayer and lipid transfer between the interface and subphase *Biophys. J.* **100** pp 1678-1687

High Performance Proactive Digital Forensics

Soltan Alharbi[1], Belaid Moa[2], Jens Weber-Jahnke[3] and Issa Traore[1]

[1] ECE Department, [2] UST Department, [3] CS Department, University of Victoria, Victoria, BC, Canada

E-mail: salharbi@uvic.ca, bmoa@uvic.ca, jens@uvic.ca, itraore@ece.uvic.ca

Abstract. With the increase in the number of digital crimes and in their sophistication, High Performance Computing (HPC) is becoming a must in Digital Forensics (DF). According to the FBI annual report, the size of data processed during the 2010 fiscal year reached 3,086 TB (compared to 2,334 TB in 2009) and the number of agencies that requested Regional Computer Forensics Laboratory assistance increasing from 689 in 2009 to 722 in 2010. Since most investigation tools are both I/O and CPU bound, the next-generation DF tools are required to be distributed and offer HPC capabilities. The need for HPC is even more evident in investigating crimes on clouds or when proactive DF analysis and on-site investigation, requiring semi-real time processing, are performed.

Although overcoming the performance challenge is a major goal in DF, as far as we know, there is almost no research on HPC-DF except for few papers. As such, in this work, we extend our work on the need of a proactive system and present a high performance automated proactive digital forensic system. The most expensive phase of the system, namely proactive analysis and detection, uses a parallel extension of the iterative z algorithm. It also implements new parallel information-based outlier detection algorithms to proactively and forensically handle suspicious activities. To analyse a large number of targets and events and continuously do so (to capture the dynamics of the system), we rely on a multi-resolution approach to explore the digital forensic space. Data set from the *Honeynet Forensic Challenge* in 2001 is used to evaluate the system from DF and HPC perspectives.

1. Introduction

1.1. Motivations

With the increase of digital crimes both in number and sophistication, High Performance Computing (HPC) is becoming a must in Digital Forensics (DF). According to the FBI annual report 2010 [3], the size of data processed during the 2010 fiscal year reached 3,086 TB (compared to 2,334 TB in 2009) and the number of agencies that requested Regional Computer Forensics Laboratory assistance increasing from 689 in 2009 to 722 in 2010. Since most investigation tools are both I/O and CPU bound, the next-generation digital forensic tools are required to be distributed and to offer HPC capabilities [10]. The need for HPC is even more evident when on-site investigation or Proactive Digital Forensic (PDF) analysis, requiring semi-real time processing, are performed. This is also the case for investigating crimes on clouds.

Although breaking the performance wall [11] is a major goal in Reactive Digital Forensic (RDF), as far as we know there is almost no research on HPC-DF except for [7, 9, 1, 13, 12, 5]. Moreover, these papers only deal with RDF. As such, our aim in this paper is to introduce HPC to PDF and contribute to the work done on HPC-RDF by

Published under licence by IOP Publishing Ltd

1. Extending the iterative z algorithm [6, 2] to different elements of DF investigation including events and targets and to take into accounts the fact that files are weighted differently. In fact, old files are usually legitimate and should be weighted more than the new ones. This weighting can be done locally (i.e. inside a directory) or globally (across all directories) and carried out using different probability distributions.
2. Parallelizing the extended algorithms using Message Passing Interface (MPI) so that the RDF and PDF analysis can be done in parallel and across distributed worker nodes.
3. Generalizing the extended algorithms to the information-based iterative z algorithms. These algorithms are introduced as a novel approach to express the outlier detection from an information theory perspective. As an example, the attribute and summary functions in the iterative z algorithm are replaced with the information and entropy functions respectively.
4. Introducing a multi-resolution approach to tackle large set of DF investigation elements for which the parallel outlier detection algorithms above may take a long time or produce many outliers. Under this approach, the outlier detection algorithms are treated as reduction operators that can be applied as many times as desired.

1.2. Related Work

In [11], Roussev *et al.* proposed a distributed DF tool and provided real-world examples where conventional investigation tools executed on a single workstation reached their performance limits. After identifying the system requirements, they designed a lightweight distributed framework to meet these requirements and implemented a prototype for it. Their experiments results were compared to Forensics ToolKit (FTK), the most popular tool in the market, and showed considerable increase in performance from hours to a few minutes.

Marziale *et al.* [7] used Graphics Processing Units (GPUs) to parallelize binary string searches offered by many DF tools. The performance results were substantially greater than the ones obtained on a multi-core CPU using threads. As an example, the time it took to carve (reassembling computer files from fragments in the absence of file system metadata) 100 GB disk image was reduced by a factor of ≈ 2.

In [9], Golden *et al.* presented an open source file carver, Scalpel, which they optimized to operate efficiently. The experiment results showed that it outperformed most of the current tools in the market. For instance, a 40 GB drive was carved using Scalpel and it exceeded Foremost by a factor of 4.

Bengtsson [1] proposed a parallel password cracker application that he ran on a Linux-based HPC cluster. He showed that by introducing such parallelism to the application, the speed increased substantially.

In [13], Roussev *et al.* emphasized that DF investigations on a large computing platform such as clouds require distributed DF tools. As a proof of concept, they implemented a distributed DF tool, MPI MapReduce (MMR). MMR demonstrated a huge increase in performance for CPU, I/O and memory bound tasks such as indexing. Wordcount application was executed under MMR and it was approximately 16 times faster than the serial execution for large files.

To cope with the new technologies and speed up the process of imaging, searching and analysing in DF, Lee *et al.* [5] suggested a new high-speed search engine and implemented it on a Tarari content processor. Their engine was validated and compared to some of the current DF tools such as EnCase. The Tarari board performed over 5 times faster than the other tools.

In [2], Carrier *et al.* adapted the iterative z algorithm, a spatial outlier algorithm [6] used to detect outliers in set of spatial features, to automate DF analysis and detect infected files. The outlier detection was done using file and directory attributes such as Modified, Accessed, and Created (MAC) times (`mtime` , `atime` , `ctime` and `size`). For single attributes, the overall results obtained on the Honeypot Challenge benchmark [4] showed a high false positive detection rate.

As such, the authors proposed using multiple attributes instead. Multiple attribute detection, however, missed some of the infected files that single attribute detection didn't.

2. Overview on Digital Forensics

Digital Forensics (DF) is defined as the ensemble of methods, tools and techniques used to collect, preserve and analyse *digital* data originated from any type of digital media involved in an incident with the purpose of extracting valid evidence for the court of law.

DF investigations are usually performed as a response to a digital crime and, as such, they are *Reactive Digital Forensic* (RDF). An RDF investigation is the traditional (or post-mortem) approach of investigating digital crimes after incidents have occurred. This involves identifying, preserving, collecting, analyzing, and generating the final report. The bottom part of Figure 1 shows the flow of the RDF investigation.

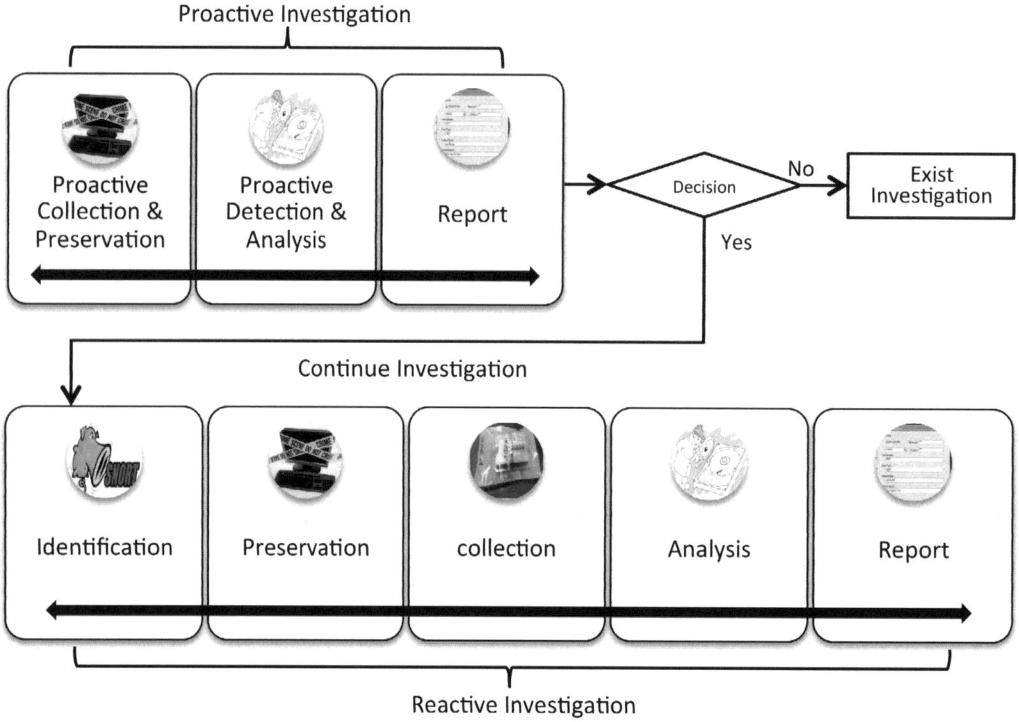

Figure 1. Proactive and Reactive Digital Forensic Framework.

Although RDF investigations are very effective, they are faced with many challenges, especially when dealing with anti-forensic incidents, volatile data and event reconstruction. To tackle these challenges, the *Proactive Digital Forensic* (PDF) [14] is required. By being proactive, DF is prepared for incidents. In fact, the PDF investigation has the ability to proactively collect data, preserve it, detect suspicious events, analyze evidence and report an incident as it occurs. The flow of PDF is depicted in Figure 1. Although there is little work done on PDF, it has many advantages over RDF such as

- Reducing the effect of anti-forensic methods.
- Providing more accurate and reliable evidence in real time.

- Promoting automation in all the phases of investigation including collection, preservation, event detection, analysis, and reporting.
- Providing reliable leads that RDF investigation can rely on.
- Saving time and money in carrying out RDF investigation.

Our focus in this paper is on the detection and analysis phase of the proactive investigation system as it is the most expensive. For more details on all of the proactive system phases, we refer the reader to [14].

3. DF and HPC

3.1. RDF and PDF Principles

HPC is needed for both reactive and proactive components of the DF investigation. This can be deduced from the five fundamental principles of computer forensics [8]:

Principle 1 Consider the entire system. This includes the user space as well as the entire kernel space, file system, network stack, and other related subsystems.

Principle 2 Trust no user and trust no policy, as we may not know what we want in advance.

Principle 3 Consider the effects of events, not just the actions that caused them, and how those effects may be altered by context and environment.

Principle 4 Consider the context in interpreting and understanding the meaning of an event.

Principle 5 Every action and every result must be processed and presented in a way that can be analysed and understood by a human forensic analyst.

These principles apply for both RDF and PDF and require considering all possible actions, targets and events as well as the context in which they occur. As such, the ultimate DF analysis is I/O-bound because keeping all the targets, events and actions requires large amount of space. The analysis is also CPU-bound as processing all the targets, events and actions and presenting the results adequately to the investigators are computationally intensive (as well as I/O intensive).

In addition to the five principles above, PDF requires, in our point of view, two extra principles:

Principle 6 Preserve the entire history of the system.

Principle 7 Perform the analysis and report the results in real-time.

Clearly, these two extra principles make PDF more demanding for HPC than RDF.

3.2. DF Multidimensional Space

In addition to the actions and events that the seven principles listed above emphasize, we introduce the notion of targets. A *target* is any resource or object related to the system under investigation including a file, memory, register, etc. We will use an *element of DF investigation* to refer to a target, an action or an event. At a time t and as shown in Figure 2, the system is in the process of executing an action that reacts to some targets and events, and produces new targets and events or modifies the existing ones. As such, to describe the dynamics of the system, one needs to know at least the state of the targets, the events generated and the actions executed at time t. Since all possible elements of investigation, each with usually many attributes with many possible values, must be considered, the space of the system dynamics is a very high multidimensional space that requires HPC for its analysis.

Being proactive implies that when many systems are involved (as is the case in networks or clouds), one has to consider not only one system but the ensemble. For this reason, we propose a simple framework in which each system is equipped with a proactive sub-agent that may carry

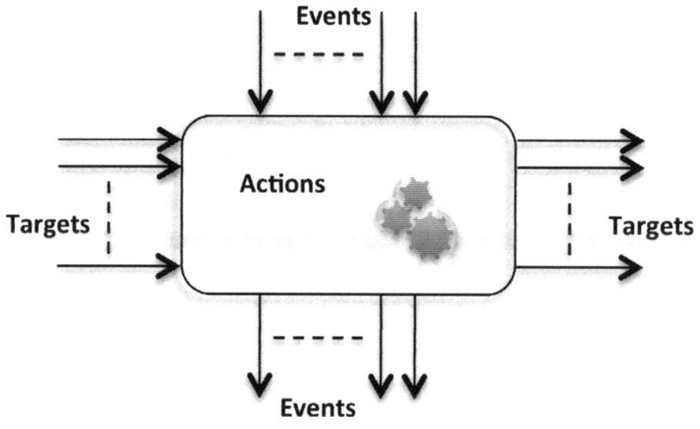

Figure 2. Relation between actions, targets and events.

out PDF but relies on a central proactive agent to do the full PDF. This agent in turn relies on the HPC cluster for storage and processing.

4. Outlier Detection for Automated DF Analysis

To handle the large number of elements of investigation and perform the PDF analysis in real-time, as the seven principles outlined, one has to automate the analysis. A promising technique for such an automation is the spatial outlier detection introduced in [2]. It is an anomaly-based technique that detects outliered targets by examining one or many of its attributes using the iterative z algorithm [6]. This algorithm finds the outlier in a set of spatial features, referred to as points, by calculating the differences between their attributes and the summary functions of their neighbours. These differences are then normalized with respect to their mean and standard deviation. When the point with the maximum normalized difference is greater than a prefixed threshold, it is marked as an outlier and its attribute is taken to be the summary function of its neighbours. The process is repeated until no outliers are found.

Carrier *et al.* adopted the iterative z algorithm to automate the DF analysis. Algorithm 1 shows how it is expressed when the elements of DF investigation are regular files and the attribute functions are the sizes of the file. In this algorithm, a regular file is denoted by ϕ_i, its size by $|\phi_i|$ and its directory by $D(\phi_i)$. The set of all files is denoted by $\mathcal{N}$ and its cardinality by $|\mathcal{N}|$.

In practice, some files should be weighted more than others. For example, new files are considered more suspicious than the old ones. As such, we introduce two kind of probability distributions: *local* and *global competitive* distributions. The latter gives the relative weights of a file with respect to all the files and the former gives its relative weights with respect to its neighbours. Algorithm 2 is the new algorithm.

If the local and the global competitive probabilities are uniform (i.e. $p^l(\phi) = 1/|N(\phi_i)|$ and $p^g(\phi_i) = 1/|\mathcal{N}|$), then we obtain Algorithm 1. But if, for example, old accessed files weight more than the new ones, then non-uniform local and the global competitive probabilities should be used as illustrated below.

Let $t_a(\phi)$ be the `atime` of a file ϕ and let t_{aref} be a reference time, which we usually take to be the latest `atime` of all the files (global `atime`) or of the files in the directory $D(\phi)$ (local `atime`). The local and the global competitive probabilities can be then taken to be

Algorithm 1 The iterative z algorithm using files as the elements of DF investigation and `size` as the attribute function.

Let θ be a threshold
Let n be $|\mathcal{N}|$
outlier=true
$\sigma = 0; \mu = 0$
for $i = 1 \rightarrow n$ **do**
  Let $N(\phi_i)$ be the set of files in $D(\phi_i)$ and $|N(\phi_i)|$ its cardinality
  Set the *attribute function* $f(\phi_i)$ to be $|\phi_i|$
  Compute the *summary function* $g(\phi_i) = \frac{1}{|N(\phi_i)|} \sum_{\phi \in N(\phi_i)} f(\phi)$
  Compute the *comparison function* $h(\phi_i) = f(\phi_i) - g(\phi_i)$
  $\mu = \mu + h(\phi_i)$
  $\sigma = \sigma + h(\phi_i)^2$
end for
$\sigma = \sqrt{\frac{\sigma}{n} - (\frac{\mu}{n})^2}$ ▷ The standard deviation
while outlier==true **do**
  outlier=false
  $\phi_q = \arg\max_\phi |\frac{h(\phi)-\mu}{\sigma}|$
  if $|\frac{h(\phi_q)-\mu}{\sigma}| \geq \theta$ **then** ▷ ϕ_q is an outlier
    Mark ϕ_q as an outlier
    $f(\phi_q) = g(\phi_q)$
    Update $g(\phi)$ and $h(\phi)$ for every ϕ in $N(\phi_q)$
    Update μ and σ
    outlier=true
  end if
end while

$$p^l(\phi_i) = \frac{(t_{aref} - t_a(\phi_i) + 1)^2}{\sum_{\phi \in N(\phi_i)} (t_{aref} - t_a(\phi) + 1)^2}$$

$$p^g(\phi_i) = \frac{(t_{aref} - t_a(\phi_i) + 1)^2}{\sum_{\phi \in \mathcal{N}} (t_{aref} - t_a(\phi) + 1)^2}$$

Depending on the value of t_{aref}, we may have p^l with either local or global `atime` and p^g with either local or global `atime`. In the case, for example, where p^l uses local `atime` and p^g uses global `atime` , the two distributions will be referred to as *global/local* `atime` distributions. The other three cases are easy to infer.

In the probabilistic iterative z algorithm, if we have

$$f(\phi_i) = -\log\left(\frac{|\phi_i|}{\sum_{\phi \in N(\phi_i)} |\phi|}\right),$$

then the attribute function f can be viewed as the information associated with each file. In addition, by taking

$$p^l_{N(\phi)}(\phi_i) = \frac{|\phi_i|}{\sum_{\phi \in N(\phi_i)} |\phi|},$$

we transform the probabilistic iterative z algorithm to the *information iterative z algorithm* shown in Algorithm 3.

Algorithm 2 The probabilistic iterative z algorithm.

Let θ be a threshold
Let n be $|\mathcal{N}|$
outlier=true
$\sigma = 0; \mu = 0$
for $i = 1 \to n$ **do**
    Let $N(\phi_i)$ be the set of files in $D(\phi_i)$ and $|N(\phi_i)|$ its cardinality
    Let $p^l_{N(\phi_i)}(\phi_i)$ be the local competitive probability of ϕ_i
    Let $p^g(\phi_i)$ be the global competitive probability of ϕ_i
    Set the *attribute function* $f(\phi_i)$ to be $|\phi_i|$
    Compute the *summary function* $g(\phi_i) = \sum_{\phi \in N(\phi_i)} p^l_{N(\phi_i)}(\phi) f(\phi)$
    Compute the *comparison function* $h(\phi_i) = f(\phi_i) - g(\phi_i)$
    $\mu = \mu + p^g(\phi_i) h(\phi_i)$
    $\sigma = \sigma + p^g(\phi_i) h(\phi_i)^2$
end for
$\sigma = \sqrt{\sigma - \mu^2}$ ▷ The standard deviation
while outlier==true **do**
    outlier=false
    $\phi_q = \arg\max_\phi |\frac{h(\phi) - \mu}{\sigma}|$
    if $|\frac{h(\phi_q) - \mu}{\sigma}| \geq \theta$ **then** ▷ ϕ_q is an outlier
        Mark ϕ_q as an outlier
        $f(\phi_q) = g(\phi_q)$
        Update $g(\phi)$ and $h(\phi)$ for every ϕ in $N(\phi_q)$
        Update μ and σ
        outlier=true
    end if
end while

In all of the previous algorithms, the comparison function $h(\phi)$ is expressed as the difference between the attribute function $f(\phi)$ and the summary function $g(\phi)$. This difference is supposed to quantify how much $f(\phi)$ is independent from $g(\phi)$. From the information theory perspective, it is more convenient to use mutual information for that. This, however, entails introducing the *cooperative* probability distribution of files in addition to their *competitive* distributions as done in Algorithms 2 and 3.

The cooperative aspect between files leads to different algorithms and we show only one for illustration: Algorithm 4. Notice how the comparison function $h(\phi)$ is expressed as the mutual information between ϕ and a virtual file with size $g(\phi)$. In addition, the inequality in line 21 has $\leq$ instead of $\geq$ used in the previous algorithms. The reason is that $h(\phi)$ expresses the dependency between files instead of their independence. We could, however, use the variation of information:

$$h(\phi_i) = H(\phi_i, \tilde{\phi_i}) - I(\phi_i, \tilde{\phi_i}),$$

and change the inequality $|\frac{h(\phi_q) - \mu}{\sigma}| \leq \theta$ to $|\frac{h(\phi_q) - \mu}{\sigma}| \geq \theta$.

In some cases, we may want to consider, the directories as independent entities as done in [2]. In these cases, one should use, in the four algorithms, the local means and the local standard deviations instead of the global ones (i.e. μ and σ).

The four algorithms can be generalized to treat any elements of investigation (instead of files only) as well as multiple attributes (instead of a single attribute, namely the `size`). They can also be adapted to PDF by running them as needed, for example, when a new element of

Algorithm 3 The information iterative z algorithm.

Let θ be a threshold
Let n be $|\mathcal{N}|$
outlier=true
$\sigma = 0; \mu = 0$
for $i = 1 \to n$ **do**
  Let $N(\phi_i)$ be the set of files in $D(\phi_i)$ and $|N(\phi_i)|$ its cardinality
  Let $p^l_{N(\phi_i)}(\phi_i)$ be the local competitive probability of ϕ_i
  Let $p^g(\phi_i)$ be the global competitive probability of ϕ_i
  Set the *attribute function* $f(\phi_i)$ to be the information $I(\phi_i) = -\log p^l_{N(\phi_i)}(\phi_i)$
  Set the *summary function* $g(\phi_i)$ to be the entropy $H(\phi_i) = \sum_{\phi \in N(\phi_i)} p^l_{N(\phi_i)}(\phi) f(\phi)$
  Compute the *comparison function* $h(\phi_i) = f(\phi_i) - g(\phi_i)$
  $\mu = \mu + p^g(\phi_i) h(\phi_i)$
  $\sigma = \sigma + p^g(\phi_i) h(\phi_i)^2$
end for
$\sigma = \sqrt{\sigma - \mu^2}$ ▷ The standard deviation
while outlier==true **do**
  outlier=false
  $\phi_q = \arg\max_\phi |\frac{h(\phi)-\mu}{\sigma}|$
  if $|\frac{h(\phi_q)-\mu}{\sigma}| \geq \theta$ **then** ▷ ϕ_q is an outlier
    Mark ϕ_q as an outlier
    $f(\phi_q) = g(\phi_q)$
    Update $g(\phi)$ and $h(\phi)$ for every ϕ in $N(\phi_q)$
    Update μ and σ
    outlier=true
  end if
end while

investigation is created or an attribute of an existing one changes. In addition, they can be run at regular intervals of time. As a remark, the previous runs of the algorithms should be used to reduce the processing by only computing the necessary functions. For example, the summary functions of investigation elements that are not neighbours of the updated elements should not be recomputed.

5. A Multi-resolution Framework for Digital Forensics

To cope with the large number of elements of DF investigation and the high dimensional space of the system, we introduce a multi-resolution approach that reduces the set of initial elements of DF investigation S_0 to a smaller one S_1. This in turn is reduced to S_2 and so on. The sequence $(S_n)_{n\in\mathbb{N}}$ obtained satisfies

$$S_0 \supseteq S_1 \supseteq \ldots \supseteq S_n \supseteq \ldots$$

A set S_{n+1} is usually the outcome of applying an operator Θ_p to S_n, that is

$$S_{n+1} = \Theta_p(S_n)$$

These kind of operators are defined next.

Definition 1. *A* DF reduction operator Θ *is an operator that maps a set S of elements of investigation to $S' = \Theta(S)$ such that $S' \subseteq S$.*

Algorithm 4 The centered mutual information iterative z algorithm.

Let θ be a threshold
Let n be $|\mathcal{N}|$
outlier=true
$\sigma = 0; \mu = 0$
for $i = 1 \to n$ **do**
  Let $N(\phi_i)$ be the set of files in $D(\phi_i)$ and $|N(\phi_i)|$ its cardinality
  Let $p^l_{N(\phi_i)}(\phi_i)$ be the local competitive probability of ϕ_i
  Let $p^g(\phi_i)$ be the global competitive probability of ϕ_i
  Set the *attribute function* $f(\phi_i)$ to be $|\phi_i|$
  Set the *summary function* $g(\phi_i) = \sum_{\phi \in N(\phi_i)} p^l_{N(\phi_i)}(\phi) f(\phi)$
  Let $p_r(\phi_i)$ be the cooperative probability of ϕ_i
  Let $\tilde{\phi}_i$ be a virtual file with size $g(\phi_i)$
  Compute the *dependency function* $h(\phi_i) = I(\phi_i, \tilde{\phi}_i)$
  $\mu = \mu + p^g(\phi_i)h(\phi_i)$
  $\sigma = \sigma + p^g(\phi_i)h(\phi_i)^2$
end for
$\sigma = \sqrt{\sigma - \mu^2}$ ▷ The standard deviation
while outlier==true **do**
  outlier=false
  $\phi_q = \arg\min_\phi |\frac{h(\phi)-\mu}{\sigma}|$
  if $|\frac{h(\phi_q)-\mu}{\sigma}| \leq \theta$ **then** ▷ ϕ_q is an outlier
    Mark ϕ_q as an outlier
    $f(\phi_q) = g(\phi_q)$
    Update $g(\phi)$ and $h(\phi)$ for every ϕ in $N(\phi_q)$
    Update μ and σ
    outlier=true
  end if
end while

Although the DF reduction operator is a new concept, it is commonly used in practice. For example, taking a snapshot of the system at a specific time t can be viewed as a DF reduction operator that reduces the elements of investigation to the files at t. Another example is generating the time line of the system in the interval $[t_i, t_e]$. It can be seen as reducing all the events and activities on the system to the ones reported by the operating system in the interval $[t_i, t_e]$. It is worth noting that a DF reduction operator can also be viewed as a projection operator that projects a DF space to a smaller one.

In our present work, the reduction operators are constructed from the spatial outlier detection algorithms. Given a set of investigation elements $\mathcal{E}$, the spatial outlier analysis reduces $\mathcal{E}$ to suspicious investigation elements using single and/or multiple attributes.

Lemma 1. *Given a sequence of DF reduction operators* $\Theta_1, \ldots, \Theta_n$, *then the composition* $\Theta_1 \circ \ldots \circ \Theta_n$ *is a DF reduction operator. In particular, if* $\Theta_1 = \Theta_2 = \ldots = \Theta_n = \Theta$, *the composition* $\Theta^n = \Theta_1 \circ \ldots \circ \Theta_n$ *is a reduction operator for every* $n \in \mathbb{N}$.

Definition 2. *A DF reduction operator* Θ *is said to* preserve V *iff* $V \subseteq \Theta(V \cup S)$ *for every DF state set* S.

In the definition above, if we take $S = \emptyset$, then we have the following lemma.

Lemma 2. *If a DF reduction operator Θ preserves V, then V is a fixed point of Θ, i.e.*

$$\Theta(V) = V.$$

Definition 3. *A DF reduction operator is* safe *if it preserves the evidence.*

Since it is undecidable to find the evidence of all kind of DF attacks, it is impossible to build a safe computational DF reduction operator.

Definition 4. *Given a DF reduction operator Θ and a set S of elements of investigation, we define the* reduction amplitude at level n *of Θ on S with respect to another set H of elements of investigation as*

$$|\Theta|_H^n(S) = 1 - \frac{|H \cap \Theta^n(S)|}{|H \cap S|}$$

In what follows, $|\Theta|^n(S)$ will be used to denote $|\Theta|_{S_0}^n(S)$, where S_0 is the initial set of investigation elements. Note that the amplitude of reduction at level n of Θ is the same as the amplitude of reduction at level 1 of Θ^n. In other words, $|\Theta^n|_H(S) = |\Theta|_H^n(S)$.

The reason for introducing the amplitude of reduction is to quantify the effectiveness of a reduction operator. For example in the Honeypot Challenge test case (see Section 6.2), when using the `mtime` attribute, the amplitude of reduction at level 1 with respect to all possible set of states S_0 is

$$|\Theta_m|(S_0) = 1 - \frac{|\Theta_m(S_0)|}{|S_0|} = 1 - \frac{8127}{20861} = 0.6$$

The `size` attribute, however, gives

$$|\Theta_s|(S_0) = 1 - \frac{|\Theta_s(S_0)|}{|S_0|} = 1 - \frac{20067}{20861} = 0.04$$

From these results we can see using the `mtime` attribute give us much better reduction than using the `size` attribute. With respect to the evidence E, however, we have

$$|\Theta_m|_E(S_0) = 1 - \frac{|E \cap \Theta_m(S_0)|}{|E \cap S_0|} = 1 - \frac{18}{60} = 0.7$$

and

$$|\Theta_s|_E(S_0) = 1 - \frac{|E \cap \Theta_s(S_0)|}{|E \cap S_0|} = 1 - \frac{58}{60} = 0.03$$

This implies that using the `mtime` attribute, the evidence was reduced compared to `size` . Therefore `size` is better when the evidence only is considered.

The goal is to have $|\Theta|$ as big as possible for the initial set S_0 but as small as possible for the evidence E. As such, we introduce the *evidence amplitude at level* n to be

$$[\Theta]^n(S) = \frac{|E \cap \Theta^n(S)|}{|E \cap S|} = 1 - |\Theta|_E^n(S)$$

It is clear that a good reduction operator should have $|\Theta|_{S_0}^n(S)$ and $[\Theta]^n(S)$ as large as possible. In other words, it should reduce S as much as possible while leaving E intact. Therefore, the perfect reduction operator is the one for which $\Theta^n(S)$ converges to $E \cap S$ as n increases.

Algorithm 5 Parallel implementation version of the spatial outlier analysis Algorithms 1, 2 and 3.

Let θ be a threshold
Group regular files according to their folders and distribute them among processes.
Each process r has a sequence of files $(\phi_i)_{i \in [a_r, b_r]}$.
Initialize the attribute functions and the probability distributions.
outlier=true
$\sigma_l = 0; \mu_l = 0$
for $i = a_r \rightarrow b_r$ **do**
    Let $N(\phi_i)$ be the set of files in $D(\phi_i)$ and $|N(\phi_i)|$ its cardinality
    Let $p^l_{N(\phi_i)}(\phi_i)$ be the local competitive probability of ϕ_i
    Let $p^g(\phi_i)$ be the global competitive probability of ϕ_i
    Compute the summary function $g(\phi_i) = \sum_{\phi \in N(\phi_i)} p^l_{N(\phi_i)}(\phi) f(\phi)$
    $\mu_l = \mu_l + p^g(\phi_i) h(\phi_i)$
    $\sigma_l = \sigma_l + p^g(\phi_i) h(\phi_i)^2$
end for
while outlier==true **do**
    outlier=false
    Reduce collectively μ_l and σ_l to μ and σ using `MPI_Allreduce` (`MPI_SUM` as its `MPI_Op`).
    $\sigma = \sqrt{\sigma - \mu^2}$
    $\phi_q = \arg\min_\phi |\frac{h(\phi) - \mu}{\sigma}|$
    Reduce collectively ϕ_q to ϕ_m using `MPI_Allreduce` with `MPI_MAXLOC` as its operation.
    Let l_m to be the location from the above `MPI_Allreduce`
    if $|\frac{h(\phi_m) - \mu}{\sigma}| \leq \theta$ **then**
        outlier=true
        if my rank is l_m **then**
            Mark ϕ_m as an outlier
            $f(\phi_m) = g(\phi_m)$
            Update $g(\phi)$ and $h(\phi)$ for every ϕ in $N(\phi_m)$
            Update μ_l and σ_l
        end if
    end if
end while

6. Implementation and Results

6.1. Implementation

We implemented parallel versions of the spatial outlier analysis Algorithms 1, 2 and 3. The implementation is based on Algorithm 5.

Since most forensic images have many directories and to reduce the amount of communications involved in computing the summary functions, we first partition the files into groups based on the directories. This partition is done using a Perl script that uses GNU Parallel to create the groups on multiple processes. To ensure a reasonable work load-balance among, say, P processes, we sort the groups ascendingly according to their number of files and distribute them in a slightly modified round-robin fashion: at each round r the process with rank $r\%P$ (remainder of r by P) is chosen to be the initial round-robin element. For each process, the name of the files it handles are written to a formatted file.

Each process then reads its file and initializes the local and global probability distributions. Different probability distributions were implemented to compare their effect on the outlier analysis as discussed below. The summary functions, the comparison functions and its partial mean and squares are then computed locally. Using `MPI_Allreduce` the mean and the standard deviation are computed. These are then used to normalized the comparison functions and compute the maximum value of the local normalized values. To get the global maximum value y_m and the rank r of the process from which it originated, we use the operation `MPI_MAXLOC` in `MPI_Allreduce` instead of `MPI_MAX`. The outlier detection is then executed by comparing y_m with a threshold θ. If y_m is greater than or equal to θ, then the process of rank r updates the necessary functions and recomputes the partial means and mean squares. All the processes then compute the global mean and the global standard deviation as outlined above. These steps are repeated until no outliers are detected (i.e. y_m is less than θ).

Since the group of files that each process handles are specified in a formatted file F, implementing the multi-resolution approach is straightforward: each time an outlier is detected, it is recorded in a formatted file similar to F. This file is read instead of F when the next multi-resolution level is desired.

Note that when the directories are assumed to be independent entities, the two MPI calls above are ignored and the code becomes a pure embarrassingly parallel one. This of course an oversimplification and it is not the case in practice as the directories are usually correlated.

6.2. Honeypot Challenge Test Case

Although our aim is to run the outlier detection algorithms in a PDF context, we needed a benchmark dataset to evaluate the effectiveness of the algorithms and to test and compare our implementation with the results reported in [2]. The comparison showed a discrepancy between our results and theirs, and hindered our efforts to implement all the algorithms and run them on a larger dataset. In fact, to validate our implementations, we had to write pure serial versions of the spatial outlier detection algorithms.

The Honeypot Challenge benchmark [4] is set of images of a computer that was attacked just after it was brought online. The images are of six partitions and the largest is the `/usr` partition (`honeypot.hda5.dd`) This partition contains a total number of 20,861 files. This number does not match the one reported by Carrier *et al.* [2] and our requests to send us the data and/or the code he used were unsuccessful as could not find neither.

The attacker used the Linux root kit version 4 (`lrk4`) to compromise the system and introduced around 60 infected files in `/usr`.

To handle the I/O bound aspect of DF, we first mounted the images on ramdisk on a cluster worker node with 24 GB of memory and 8 cores with read only access. We then run the serial and parallel versions of different algorithms to make sure that our parallel implementation results match the serial ones. All the parallel algorithms, including the ones using the multi-resolution technique, completed in less than 2 minutes. On a Dual core MacBook Pro with 4 GB of memory, it took almost 12 minutes. Although the dataset used is very small, we obtained a speedup of 6.

The analysis was done for many attributes including `size` , `mtime` , `atime` , `ctime` and `inode` number, for different global and local probability distributions such as uniform, `mtime` , `atime` and `ctime` distributions, and for the information iterative z algorithm. Some of the results obtained are presented in Figures 3, 4, 5, 7 and 8.

A brief analysis of the results is as follows:

- The results we obtained for `size` , `ctime` and `mtime` with uniform distributions do not match that of Carrier *et al.* reported; our results have higher false positives than Carrier *et al.* As mentioned previously, we were unsuccessful to get neither the dataset nor the code they used to reproduce their results. Our pure serial implementations, however, confirmed the accuracy of our results.
- The reduction amplitude for `mtime` and `inode` are higher than that of `size` , `atime` and `ctime` .

$\theta = 2$, Prob. Dist. = Uniform						
Multi-resolution Level	`size`		`mtime`		`inode`	
	Reduction Amp.(%)	Evidence Amp.(%)	Reduction Amp.(%)	Evidence Amp.(%)	Reduction Amp.(%)	Evidence Amp.(%)
1	4	97	60	30	4	83
2	4	97	65	28	7	68
3	5	90	91	10	8	65
4	6	80	93	8	8	65
5	6	80	93	8	8	65

Figure 3. Comparing `size`, `mtime` and `inode` under uniform distributions.

$\theta = 2$, Prob. Dist. = Global/Local `mtime`						
Multi-resolution level	`size`		`mtime`		`inode`	
	Reduction Amp.(%)	Evidence Amp.(%)	Reduction Amp.(%)	Evidence Amp.(%)	Reduction Amp.(%)	Evidence Amp.(%)
1	3	88	55	27	2	88
2	6	80	67	18	3	80
3	43	72	74	15	3	78
4	76	52	88	10	4	78
5	78	50	89	10	4	78

Figure 4. Comparing `size`, `mtime` and `inode` under global/local `mtime` distributions.

$\theta = 2$, `mtime`				
Multi-resolution level	Prob. Dist. Uniform		Prob. Dist. Global/Global `atime`	
	Reduction Amp.(%)	Evidence Amp.(%)	Reduction Amp.(%)	Evidence Amp.(%)
1	61	30	61	27
2	65	28	65	22
3	91	10	66	22
4	93	8	78	17
5	93	8	79	17

Figure 5. Comparing uniform and global/global `atime` distributions under `mtime`.

- As the tables in Figures 6 and 7, the `mtime` , `ctime` and `inode` probability distributions were effective in increasing the evidence amplitude even though user activity on the system was very little. This increase, however, decreased the reduction amplitude.
- The information-based outlier detection using `mtime` outperformed the rest of algorithms in most cases.
- Global/local distribution can help in increasing the reduction amplitude as the multi-resolution level increases as shown for example in Figure 4.

7. Conclusions and Future work

In this paper, we argued from the first five principles of DF that HPC computing is required to carry out automated DF analysis. In addition to extending the principles of RDF to PDF,

$\theta = 2$, Directories as Independent Entities				
Local Prob. Dist.	`size`		`inode`	
	Reduction Amp.(%)	Evidence Amp.(%)	Reduction Amp.(%)	Evidence Amp.(%)
Uniform	72	15	94	3
`atime`	74	15	81	3
`ctime`	72	15	94	3
`mtime`	68	30	77	20
`inode`	62	10	36	55

Figure 6. Comparing `size` and `inode` when the directories are independent entities.

$\theta = 2$, Directories as independent entities, Processed all files and directories				
Local Prob. Dist.	`inode`		`mtime`	
	Reduction Amp.(%)	Evidence Amp.(%)	Reduction Amp.(%)	Evidence Amp.(%)
Uniform	91	12	91	20
`atime`	86	13	87	20
`ctime`	83	50	70	42
`mtime`	76	30	76	42
`inode`	38	77	83	55

Figure 7. Comparing `inode` and `mtime` when both directories and files are used, and the directories are independent entities.

$\theta = 2$, Information-based detection using `mtime`				
Multi-resolution level	Prob. Dist. Global/Global `mtime`		Prob. Dist. Local/Local `mtime`	
	Reduction Amp.(%)	Evidence Amp.(%)	Reduction Amp.(%)	Evidence Amp.(%)
1	59	43	46	35
2	66	38	56	22
3	91	12	59	20
4	92	10	60	20
5	93	10	62	20

Figure 8. Comparing global/global and local/local `mtime` distributions under information iterative z algorithm using `mtime`.

we parallelized and extended the iterative z algorithm, which is used to detect the suspicious elements of DF investigation. We then generalized our extended algorithm to information and mutual information-based algorithms. These algorithms are then incorporated, as DF reduction operators, into a multi-resolution framework that permit us to iteratively reduce the number of outliers.

Although the performance of our implementation is so far encouraging (the results of the analysis were obtained in less than 2 minutes), we need to test our algorithms on large datasets. The dataset used is very small and cannot be used as benchmark for evaluating the performance of our system. Moreover, the algorithms were tested in an RDF context and should need to be adapted and incorporated to PDF.

Since the information iterative z algorithm outperformed the extended iterative z algorithms in most cases, we plan to implement the rest of information-based iterative z algorithms including Algorithm 4. Given that DF reduction operators can be composed and each algorithm behaves as a reduction operator, deciding the order of applying them is a challenging task that we will be addressing. Moreover, the outlier detection algorithms implemented so far use single attributes to compute the summary functions and the probability distributions. Our next future work is to extend them to multiple attributes as done in [2, 6].

Acknowledgements

This research has been enabled by the use of computing resources provided by WestGrid and Compute/Calcul Canada. We also acknowledge the generous support of Taif University and the Center of Excellence in Information Assurance (CoEIA) Saudi Arabia.

References

[1] J. Bengtsson. Parallel password cracker: A feasibility study of using linux clustering technique in computer forensics. In *Digital Forensics and Incident Analysis, 2007. WDFIA 2007. Second International Workshop on*, pages 75–82. IEEE, 2007.

[2] B.D. Carrier and E.H. Spafford. Automated digital evidence target definition using outlier analysis and existing evidence. In *Proceedings of the 2005 Digital Forensic Research Workshop (DFRWS)*. Citeseer, 2005.

[3] Federal Bureau of Investigation. Program annual report for fiscal year 2010. `http://www.rcfl.gov/downloads/documents/RCFL_Nat_Annual10.pdf`, 2010.

[4] Honeynet Project. Forensic challenge. `http://old.honeynet.org/challenge/index.html`, 2001.

[5] J. Lee, S. Un, and D. Hong. High-speed search using tarari content processor in digital forensics. *digital investigation*, 5:S91–S95, 2008.

[6] C.T. Lu, D. Chen, and Y. Kou. Algorithms for spatial outlier detection. In *Data Mining, 2003. ICDM 2003. Third IEEE International Conference on*, pages 597–600. IEEE, 2003.

[7] L. Marziale, G.G. Richard III, and V. Roussev. Massive threading: Using gpus to increase the performance of digital forensics tools. *digital investigation*, 4:73–81, 2007.

[8] S. Peisert, S. Karin, M. Bishop, and K. Marzullo. Principles-driven forensic analysis. In *Proceedings of the 2005 workshop on New security paradigms*, pages 85–93. ACM, 2005.

[9] G.G. Richard III and V. Roussev. Scalpel: A frugal, high performance file carver. In *Proceedings of the 2005 digital forensics research workshop (DFRWS 2005)*, 2005.

[10] G.G. Richard III and V. Roussev. Next-generation digital forensics. *Communications of the ACM*, 49(2):76–80, 2006.

[11] V. Roussev and G.G. Richard III. Breaking the performance wall: The case for distributed digital forensics. In *Proceedings of the 2004 digital forensics research workshop (DFRWS 2004)*, 2004.

[12] V. Roussev, L. Wang, G.G.R. Iii, and L. Marziale. Mmr: A platform for large-scale forensic computing. 2009.

[13] V. Roussev, L. Wang, G. Richard, and L. Marziale. A cloud computing platform for large-scale forensic computing. *Advances in Digital Forensics V*, pages 201–214, 2009.

[14] S.A. Soltan Alharbi, J.W.J. Jens Weber-Jahnke, and I.T. Issa Traore. The proactive and reactive digital forensics investigation process: A systematic literature review. *International Journal of Security and Its Applications*, 5(4):59–72, 2011.

Hybrid resource provisioning for clouds

Mahfuzur Rahman and Peter Graham

Dept. of Computer Science, University of Manitoba, Winnipeg, MB, Canada R3T 2N2

E-mail: {mahfuz,pgraham}@cs.umanitoba.ca

Abstract. Flexible resource provisioning, the assignment of virtual machines (VMs) to physical machine, is a key requirement for cloud computing. To achieve "provisioning elasticity", the cloud needs to manage its available resources on demand. A-priori, static, VM provisioning introduces no runtime overhead but fails to deal with unanticipated changes in resource demands. Dynamic provisioning addresses this problem but introduces runtime overhead. To reduce VM management overhead so more useful work can be done and to also avoid sub-optimal provisioning we propose a hybrid approach that combines static and dynamic provisioning. The idea is to adapt a good initial static placement of VMs in response to evolving load characteristics, using live migration, as long as the overhead of doing so is low and the effectiveness is high. When this is no longer so, we trigger a revised static placement. (Thus, we are essentially applying local multi-objective optimization to tune a global optimization with reduced overhead.) This approach requires a complicated migration decision algorithm based on current and predicted:future workloads, power consumptions and memory usage in the host machines as well as network burst characteristics for the various possible VM multiplexings (combinations of VMs on a host). A further challenge is to identify those characteristics of the dynamic provisioning that should trigger static re-provisioning.

1. Introduction

The growth in popularity of clouds for providing IT services continues at a rapid pace. Cloud-based approaches are attractive because enterprises do not need to purchase, maintain and support their own computing infrastructure which often represents a significant and sometime hard to predict operational cost. Further, the large scale of cloud infrastructure (tens of thousands of cores and beyond) allows cloud providers to optimize the efficiency of the systems, via the use of virtual machines (VMs) to permit effective sharing, so they can offer services more cost-effectively than end users could. This makes the use of clouds price-attractive as well.

End users and cloud providers typically enter into contracts, quantified as service level agreements (SLAs), to describe the expected requirements, in terms of computing, etc. capacity, required for each user application. Cloud providers use this information to determine the amount of infrastructure they need to purchase and later allocate to users when their applications run as well as the cost to the users. This allocation is done by "provisioning" algorithms that pack virtual machines running end-user applications into physical cloud infrastructure (compute nodes) that the cloud provides.

For applications that have predictable and generally unchanging demands/workloads, it is possible to use "static provisioning" effectively. In this approach VMs are carefully packed into clouds as they are brought online and continue to execute effectively for potentially very long periods of time on the selected nodes. Using static partitioning, the cost of making the placement decision is incurred once, before the user's application is running, and therefore may safely incur greater overhead to achieve a better result without negatively impacting the end users. In cases where demand by applications may change or vary, "dynamic provisioning" techniques have been suggested whereby VMs may be migrated on-the-fly to new compute nodes within the cloud. This allows the system to adapt to changing conditions but at the cost of runtime overhead and potential execution delays that may negatively affect end users. To minimize this overhead, restrictions on the complexity of provisioning algorithms are necessary.

To effectively address scenarios with changing load (e.g. web servers hosting hot content) we propose a hybrid provisioning approach that starts with a good static placement that can be tuned by dynamic re-provisioning as long as doing so is effective and not excessively costly. As the overhead of VM migrations increases, re-provisioning of course becomes less attractive. At that point, we propose to incur the greater cost of some form of static re-provisioning of potentially all running VMs based on the newly available demand information to return the system to a state of efficient sharing and execution. This can be done concurrently with the ongoing execution of VMs in the cloud as long as there are available idle resources – a likely situation or one that can be planned for.

To implement and assess hybrid provisioning strategies, it is necessary to have both a testbed environment for verifying the correctness and small-scale performance of developed algorithms as well as a simulation system to assess larger-scale characteristics. In this paper, we describe our extensions to the Eucalyptus open source cloud system to support hybrid provisioning, we discuss our hybrid algorithm framework and then briefly discuss our planned evaluation strategy using CloudSim based on the framework and results from our modified version of Eucalyptus.

The rest of this paper is organized as follows. Section 2 briefly reviews work related to this paper. Section 3 states the problem to be solved and Section 4 describes the proposed solution. Our prototype, framework and planned assessment strategy are discussed in Section 5. Finally, Section 6 presents our conclusions and our planned directions for future work.

2. Related Work

Cloud computing [1–3] is based on large-scale computing infrastructure and introduces an alternative way to efficiently deliver IT services. Clouds can offer hardware resources (Infrastructure-as-a-Service/IaaS), software platforms (Platform-as-a-Servce/PaaS) or hosted applications (Software-as-a-Service/SaaS) on an on-demand basis. Clouds consist of a huge number of LAN-interconnected physical machines supporting carefully assigned virtual machines that ultimately run the end-user applications that are accessed remotely by clients.

2.1. Provisioning in Clouds

In clouds, application developers, who are sometimes referred to as SaaS (Software as a Service) providers or cloud clients, pay to allocate as much computational resources; memory space, disk storage, network bandwidth, etc. as they need to host their applications based on their current and/or expected load. Depending on SaaS providers' requirements, cloud infrastructure providers allocate the necessary resources, in the form of Virtual Machines (VMs). Efficient resource utilization is a key requirement in clouds from the provider's perspective and VMs are used in cloud computing to enable efficient resource provisioning as well as enhanced security. Cloud-based data centers normally provide a huge amount of hardware resources (CPU, memory, network capacity, etc.) in the form of a large number of compute nodes networked together to form a massive, efficiently sharable computing resource. These physical resources in the cloud are then made available to different VMs to run applications. VM provisioning helps ensure efficient application execution, load balancing and proactive failure handling, and thus can increase the reliability and efficiency of the whole cloud system and the applications running in it. VMs may be assigned to the physical machines in a cloud using either static or dynamic VM provisioning,

In static VM provisioning, the required hardware resources are selected as the VMs are requested based on agreed-to Service Level Agreements (SLAs) but in the case of dynamic VM provisioning, the resources used can be changed dynamically to handle unexpected fluctuations in workloads. Simple static algorithms track available resources and allocate VMs to physical machines using techniques designed to achieve certain goals. Techniques such as Round Robin (where VMs are evenly spread over the physical machines to achieve load balance and, presumably, good performance) and Greedy (where the goal is to pack VMs onto the fewest physical machines to achieve low power consumption since unused machines may be powered down) are examples of such algorithms. Of course, for both static and dynamic provisioning, "sizing" of the required resources must be done to determine the appropriate hardware resources needed. Meng et al. [4] introduced joint VM sizing which can aggregate the total capacity requirements of multiple VMs. Joint VM sizing is useful when *multiple* VMs are hosted in the same physical host and the unused resources, identified by previous resource usage patterns and/or future usage prediction, of one VM can be allocated to other VM(s). Total capacity measurement is used to select the appropriate VMs to host in a particular physical machine. In static VM provisioning, the resources that are allocated to a VM may not be fully utilized and in that case VM multiplexing helps to improve utilization of the hardware resources. As part of their work, Meng et al. also developed a workload forecasting model to determine compatible VMs that can be consolidated on a single physical machine without violating any of the VM's SLAs. Fito et al. [5] developed related criteria for cloud-based SLAs and a model to monitor SLA satisfaction in clouds.

In dynamic VM provisioning, additional required resources can be dynamically allocated for a VM when its resource requirement exceeds the currently allocated resources. Also resources can be de-allocated when a VM doesn't need them. Kim et al. [6] proposed a prediction-based dynamic VM provisioning model based on the previous resource usage patterns of a VM. They introduced pattern based provisioning (where additional resources are provisioned depending on resource usage patterns) to avoid the negative affects of threshold based provisioning (where additional resources are provisioned only when a certain threshold of resource use has already been crossed) which sometimes leads to frequent and inappropriate provisioning requests. Rolia et al. [7] also used historical resource use information to predict future resource requirements for applications. Though previous use information may not always provide accurate prediction, especially for web-based interactive applications and in the case of shared virtualized infrastructure where resource availability also depends on the requirements of co-hosted applications and their priorities, this approach can decrease the number of active hosts/nodes thereby decreasing cloud resource requirements and power consumption.

2.2. Virtual Machine Migration

Sometimes the resource requirements of a VM may exceed the capacity of the current host. In that case, the VM needs to be migrated to another, more suitable host. (This is how dynamic VM provisioning is effected.) VMs can also be migrated off physical machines when the current host requires maintenance or to reduce power consumption by consolidating VMs into a smaller number of hosts so the unused hosts can be shut down. Clark et al. [8] described "live" migration of a VM across physical hosts where execution carries on either in the source or in the destination host during the transfer of memory contents. At a certain point, VM migration needs to stop execution in the VM at the source host and to start execution at the destination. The relatively short time between these events is when no progress can be made. In VM migration, any such downtime is important to minimize since it may degrade the system's ability to meet the relevant VM SLAs. In VM migration, CPU and network overheads may also degrade application performance. In spite of this, VM migration is very useful to deal with runtime overload situations, especially in the case of sustained overload [9]. Hines et al. [10]

compare pre-copy and post-copy based live migration of VMs. In pre-copy based live migration, the CPU state transfer takes place after the memory transfer phase and in post-copy based live migration, the CPU state transfer takes place before the memory transfer phase.

2.3. Assessing Cloud Systems

Assessment of cloud systems (including VM provisioning techniques) is complicated by the scale of cloud infrastructure. Few researchers have access to large-scale cloud facilities or data from them that they can use to do assessments. Small scale prototype systems can and should be used to demonstrate algorithm correctness but such systems cannot be used to assess scale-related effects. To address this shortcoming, some form of simulation is therefore also required. Most cloud systems software is commercial and hence proprietary and unavailable to academic researchers. An exception is the Eucalyptus [11] open source cloud system which provides a basic but largely complete cloud testbed environment supporting cloud infrastructure consisting of clusters containing compute nodes running a variety of operating systems under a number of different virtual machine management packages. A number of simulation systems have also been developed for assessing cloud-based systems. Perhaps the most common and widely used of these is the Cloudsim [12, 13] simulator which supports the simulation of both the system and behaviour of cloud components including VMs, resource provisioning policies, etc. Of importance to the work described in this paper, it also provides custom interfaces for implementing new provisioning techniques. An issue related to the simulation of clouds is the generation of cloud workloads to use with simulation systems. As many cloud applications are web-based, the use of tools such as Olio [14] and the Faban web load generator [14, 15] to create reasonably realistic synthetic workloads is common.

3. The Problem

A resource provisioning "bottleneck" occurs when a cloud cannot provide the needed resources to applications when those applications need them (e.g. in response to changing load conditions). This means that SLAs may not be met in the face of increased load. Static VM provisioning, the most common current strategy, introduces no runtime VM management overhead but fails to address such resource bottleneck problems. On the other hand, dynamic VM provisioning can address or, possibly, even proactively avoid resource provisioning bottlenecks, but it introduces additional VM management overhead due to the need to predict future application workloads, and then allocate VMs to appropriate physical machines at runtime. Ideally, a provisioning solution would be able to offer the adaptability of dynamic techniques but with less VM management overhead than is currently the case using dynamic provisioning. To accomplish this, a different approach is required.

4. The Proposed Solution

The challenge of dynamic provisioning is to ensure that the benefit of any migration significantly outweighs the overhead of doing so. We propose to combine the best features of both static and dynamic VM provisioning via a hybrid approach where dynamic tuning carefully refines a good initial static placement and, if necessary, may eventually trigger a revised static placement. At least two challenges must be met in any such hybrid provisioning technique. First, it must be possible to determine when the migration of a VM should take place to avoid resource provisioning bottlenecks and to provide efficient and cost effective use of physical machines in the cloud. Second, it must be possible to determine whether or not a proposed migration will provide ongoing performance benefit. Meeting these challenges will require an accurate prediction model that will depend on both the current and predicted future workload of the current host, current and predicted future workload of other hosts, power consumption (if the selected destination host needs to be powered on), memory usage patterns, network burst characteristics, various multiplexing possibilities with other VMs, etc.

By combining static and dynamic provisioning we expect to create a system where a good, inexpensive static provisioning can be cost-effectively "tuned", when necessary, by a careful and hence lower overhead dynamic migration-based technique thereby providing adaptation to unanticipated changes in cloud workload. Further, we expect to be able to identify characteristics of the dynamic provisioning done that can be used to trigger static re-provisioning to limit the need for future dynamic provisioning and associated overhead (such static re-provisioning can be done, up to the point of migrating VMs, entirely in parallel with ongoing operations in the cloud) thereby limiting the impact on VM performance and SLA conformance. Finally, we plan to develop a technique for selecting the order of VM migration to achieve a selected re-provisioning that will minimize disruption to computations taking place in the cloud.

5. Our Prototype and Framework

To assess the effectiveness of hybrid algorithms for resource provisioning in clouds we have to use a mixture of prototype implementation and simulation. This paper addresses the development of a suitable testbed prototype based on an extension of the Eucalyptus open source cloud system that is used to ensure algorithm correctness. We begin by describing the structure of Eucalyptus and then the changes that were necessary to make to it so it could serve as a platform for testing hybrid provisioning algorithms.

5.1. Eucalyptus

Eucalyptus implements IaaS (Infrastructure as a Service) level cloud computing functionality which is very similar to Amazon's EC2 [16] and can be easily extended due to its simple organization and modular design. There are three major high level components in Eucalyptus: the Node Controller (NC), the Cluster Controller (CC) and the Cloud Controller (CLC) as shown in Figure 1.

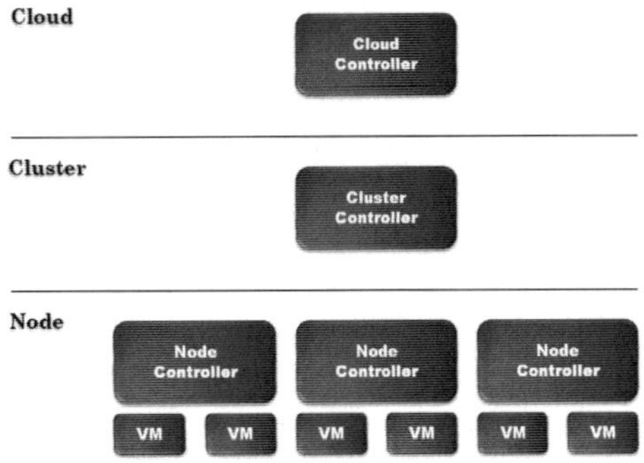

Figure 1. Different Levels in Eucalyptus (adapted from [11])

Node Controllers (NCs) are installed on each physical node/host machine that is part of the Eucalyptus cloud infrastructure. Each NC makes the physical resources (i.e. computing core(s), memory, storage etc.) of the associated node available to the Eucalyptus cloud. When any new node is attached to a cloud, the Cloud Controller (CLC) can then allocate and manage its physical resources in the form of VM instances via the NC of that particular node. Nodes, having the same type of hypervisor, are arranged under a cluster where each cluster contains

Journal of Physics: Conference Series **385** (2012) 012004 doi:10.1088/1742-6596/385/1/012004

a Cluster Controller (CC). The single CLC communicates with NCs through each particular cluster's CC. Each NC can query, control, allocate and/or manage the host operating system, the hypervisor of the node, VM instance(s) on the node, etc. upon request from the CLC.

Cluster Controllers (CCs) act as intermediaries between the CLC and the NC(s) registered in the corresponding cluster. Each CC gathers VM instance information from its associated nodes. Each CC also handles requests for new VM instance creation (from the CLC) by determining a node that has enough free resources to satisfy the new VM instance's resource requirements (i.e. it does the VM provisioning). Additionally, each CC passes-thru VM instance termination or reboot requests from CLC to the relevant node's NC. The interfaces that maintain the communication between CC and NC is described using WSDL/SOAP.

The Cloud Controller (CLC) is the management entry point and major decision making component in Eucalyptus. The CLC is accessed using a number of web-services which can be categorized into: a resource management service, a data service and an interface service. Resource management services interact with CC(s) to allocate and de-allocate physical resources and to create and control VM instances. The data services create, integrate, modify and contain information related to users, systems and data storage. This information normally includes key-pairs, security groups, network definitions, management and monitoring information of VM instances, etc. The interface services are provided to allow communication with internal Eucalyptus objects using standard WS-security policies and to initiate high-level system management tasks.

At present there are three static VM provisioning techniques implemented (by the CCs) in Eucalyptus: the Round Robin, Greedy and PowerSave algorithms. The Round Robin algorithm allocates new VMs to the host machines in a circular fashion distributing the load across the available nodes. The Round Robin algorithm maintains a list of host machines and also a counter for identifying the next available host machine in that list. The counter is incremented modulo the number of nodes after each new VM allocation to point to the next host machine in the list where the next VM allocation will take place. The Round Robin algorithm can prevent starvation as it considers each host machine equally and it is very good for balancing load (i.e. equal distribution of total workload among the host machines in the cloud). The Greedy algorithm maintains a sequence of host machines with available capacity and when a new VM needs to be allocated, the Greedy algorithm uses that sequence to find possible VM allocation locations. The Greedy algorithm allocates the new VM to the first host machine meeting the needs of the new VM thereby packing VMs onto as few nodes as possible. The Greedy algorithm provides locally optimal allocation but often fails to achieve the best overall optimal solution as it always considers the first possible solution without considering other possible allocation locations. Though the Greedy algorithm cannot balance the total VM workload among the host machines, it is faster because of less allocation workload. The PowerSave algorithm enhances the Greedy algorithm by putting host machines to sleep when they are not running any VM and reawakens a host machine when it is needed for allocating new VM(s). By shutting down the unused host machines, the PowerSave algorithm can reduce overall power consumptions but it decreases the speed of allocation due to the need to reawaken a node if no capacity exists on other nodes.

5.2. Implemented Changes to Eucalyptus

In Eucalyptus, the CLC contains 'euca2ools' which is used to query and display information about currently running VM instances in the cloud. The information produced provides VM instances' ID along with their IP addresses and the relevant cluster's name. This information can be used to access any particular VM instance, through its IP address, to terminate or to reboot the VM instance. But the query result doesn't include information related to the node where the VM instance is currently running. The query result also doesn't provide any resource usage information for VM instances. Both of these pieces of information are required to support a cloud-wide hybrid provisioning approach. We have therefore enhanced Eucalyptus so that a query result now also provides the corresponding nodes' performance information, considering available and used resources, along with their hosted VM instances' performance information, also considering allocated and used resources. We have also modified the data structures maintained in the CLC, CCs, and NCs and also the message passing parameters among them to permit collection and storage of the required performance information. This performance information provides the basis upon which we can identify over-utilized and under-utilized nodes and VM instances in the cloud that may need to be migrated as part of a hybrid provisioning technique.

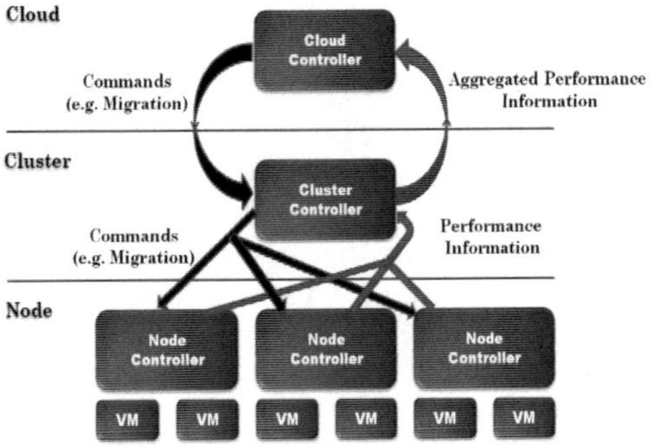

Figure 2. Modified Communications (blue arrows) and Control (brown arrows) in our Modified Eucalyptus (compare to Figure 1).

Currently in Eucalyptus the VM provisioning sub-module occurs in each CC and provides different provisioning algorithms (such as Round Robin, Greedy, etc.) to identify a suitable node to place each new VM instance on as it is created. This obviously does not permit provisioning across the cloud as a whole and also prevents the possibility of migration across clusters in the cloud (due to lack of overall placement information) thereby limiting choices in re-provisioning. We have therefore moved the VM provisioning sub-module to the CLC level instead of at the CC level. This rearrangement allows us to allocate new VM instances or to migrate existing VM instances to a node across clusters in the cloud. In the CLC, we have also introduced a new communication interface permitting migration decisions to be delivered to the affected NCs. To do this, each CC was modified to pass-thru VM instance migration requests from the CLC to the relevant node's NC and we have made each NC capable of directing its hypervisor to perform live migration of a VM instance. (Refer to the control communication paths on the left side of Figure 2). We have also enhanced the NCs to communicate with their underlying hypervisors to query and collect VM instances' performance information and we have modified the CCs so that when they gather VM instance information from their associated nodes they also collect the additional resource usage information of the nodes and the VM instances running on them. Additionally, each CC summarizes the information collected and forwards the information up to the CLC, as shown in the data communication paths on the right side of Figure 2. Finally, we have made it so each NC periodically refreshes and updates the CCs about the resource use by their nodes and VM instances.

All of these changes have been implemented in Eucalyptus and deployed on a small scale cloud testbed system running the modified Eucalyptus.

5.3. Framework for Hybrid Provisioning Algorithms

Given the pre-requisite ability to collect performance information on VM instances and nodes, the first component of a hybrid provisioning scheme, assuming an existing static placement mechanism (as is available in Eucalyptus), is a mechanism to decide when and which VMs should be migrated. As discussed earlier, this must include a complex heuristic assessing the costs and benefits of a VM migration which includes many factors. A key challenge to implementing such a heuristic is that migration decisions must not be made based on transient changes in system load and hence performance. Reacting too quickly to changes in load may result in unnecessary and ultimately non-beneficial VM migrations. An appropriate heuristic must be able to ensure that the benefits of migration will continue to be effective over time. Ultimately, the inability to meet this criteria suggests that a static re-balancing (i.e. re-running a static provisioning algorithm for *all* existing VMs) may be necessary. We now sketch our initial framework for supporting such heuristic algorithms. Our framework for migrating a single VM based on a threshold being exceeded over some period of history is shown in Algorithm 1. Naturally, use of this framework provides for local but not global optimization in re-provisioning.

To implement the framework, each CC collects the performance information from each of its registered nodes and the VMs running in them using `getNodeResourceUsageHistories()` [line 1, Algorithm 1]. The collected information is forwarded to the CLC by each CC via `getClusterResourceUsageHistories()` [line 2]. After aggregating and sorting the performance information from the CCs according to load/utilization [lines 3–4], the CLC iterates through the usage records corresponding to over-utilized nodes [lines 5-6]. For each such node it generates a set of potential `underUtilizedNodes` that can support the originally specified (i.e. SLA) requirements (`r.resourceRequirements`) via a call to `findUnderUtilizedNodes()` [line 7]. The nodes generated are used [line 8] together with the history of actual resource usage (`r.resourceUsageHistory`) to create a list of `selectedNodes` that could be good migration targets. Information in `r.resourceUsageHistory` permits assessment of past variation in use from the expected use (`r.resourceRequirements`) which is used in predicting future use. The list of potential migration targets is ordered by the expected benefit of migration to that node. The function `benefitAnalysis` that produces the list considers such additional factors as the predicted longer-term network bandwidth and latency available, the cost of memory state migration, file migration and network re-configuration as well as the benefits of power savings etc. Migration is only done if the benefit of the migration to the `selectedNode` exceeds another threshold (`thresholdBenefit`) [lines 10-11].

In our framework, resource requirements, resource usage and threshold information are all compound representations encompassing such factors as CPU core, memory, storage, network, etc. need, use and limits, respectively. While the pseudo-code in Algorithm 1, presents operations on these values as simple comparisons, they are not. In different situations, the operations required may differ in behaviour to ensure appropriate outcomes. Generally, comparisons are done component-wise (e.g. CPU need vs. CPU available, memory need vs. memory available, etc.). Simple absolute comparisons of components, however, may be inappropriate in some cases so a "fuzzy" form of comparison can be done where the degree of flexibility may vary between component values during comparisons. (E.g. The need for CPU capacity may be harder than for expected available network capacity.) Similarly, the thresholds referred to in the algorithm (`thresholdResourceUsage` and `thresholdBenefit`) appear to be absolute and arbitrary. Initially these values will be absolute but must be carefully chosen experimentally. In the future, we will make the setting of these thresholds dynamic based on sampled system behaviour to adapt to changing global conditions in the cloud system and thereby improve the overall performance benefit of the migrations selected.

Hybrid Algorithm Framework: Threshold-Driven Dynamic Provisioning (Single VM)

```
clusterHistories ← getNodeResourceUsageHistories( )
cloudHistory ← getClusterResourceUsageHistories(clusterHistories)
usageRecords ← extractRecords(cloudHistory)
usageRecords ← sortRecordsByUsage(usageRecords)
foreach record r in usageRecords do
    if r.resourceUsage ≥ thresholdResourceUsage then
        underUtilizedNodes ← findUnderUtilizedNodes ( r.resourceRequirements )
        selectedNodes ← benefitAnalysis(underUtilizedNodes, r.resourceUsageHistory)
        selectedNode ← first element in selectedNodes
        if selectedNode.benefit ≥ thresholdBenefit then
            performMigration ( r.InstanceId, selectedNode)
        end if
    else
        exit // no remaining records exceed threshold
    end if
end for
```

Algorithm 1.

5.4. Status and Planned Evaluation Strategy

At this point we have made all the changes to Eucalyptus and tested their functionality. We are refining the details of the framework described and should be implementing and testing the first of a family of algorithms shortly. Once we have working algorithms, it will be necessary to evaluate their effectiveness against static provisioning alone. Intuitively, as the workload variation increases over time, the effectiveness of rigid static provisioning should decrease. At this point, application of dynamic re-provisioning, as described, should improve effectiveness. If the changes in workload are gradual dynamic re-provisioning alone should remain effective for some time. However, if the changes are more significant and/or as changes compound over time we expect the overhead of dynamic re-provisioning to grow to the point where it represents a significant drain on the cloud system's capacity. It is at this point that we will trigger a static repartitioning (i.e. the use of a static algorithm that will look at a large-scale re-provisioning for all running VMs). While the cost of such a re-partitioning will be high (and will, necessarily, have to be done in a phased-approach to avoid disruption of cloud services) the end result should limit the need for ongoing migrations and their associated overhead. This leads directly to the question of "How we will assess the effectiveness of our hybrid provisioning algorithms?"

Ultimately, any evaluation must focus on how much the proposed integration of static and dynamic VM provisioning improves overall resource utilization in a cloud while maintaining SLA commitments. We will conduct experiments to measure capacity savings with and without using hybrid techniques. We will first test each of our implemented algorithms in our small scale modified Eucalyptus cloud testbed environment to assess correctness and accuracy, in terms of selecting the appropriate VMs to migrate, and to collect actual performance data (e.g. live migration costs). We will then do further, scale-related experiments in a simulated cloud environment based on Cloudsim.

We will use Cloudsim [12, 13] to create a much larger simulated cloud environment for our assessment of scale related effects. From our initial evaluation it seems that the Cloudsim simulator will support the straightforward addition of the necessary hybrid provisioning policies and VM migration behaviour. As we do not currently have access to cloud system trace data we plan to follow common practice and use Olio [14] along with the Faban load generator [14, 15] to create reasonably realistic synthetic workloads to drive our simulations. Olio allows the creation of sample Web 2.0 applications to evaluate the performance of web based cloud applications. Faban can be used along with Olio to create a Markov-chain based workload for the VMs (this approach was previously used in [17]). We will also continue looking for suitable trace data from real world cloud implementations which would help to provide validation of our simulation experiments, if obtained.

Using our simulated cloud environment, we will assess the effectiveness and duration of migration in a large-scale, realistic cloud scenario based on data collected from the prototype implementation. We will also analyze trends in the effectiveness of migration over time to help determine when and under what conditions, static re-provisioning may be necessary. Our performance metric will be the number of VM migrations eliminated using our approach. Fewer VM migrations reduces overhead and thus improves overall system performance. We will need to assess the total number of VM migrations within a time period for different workloads, considering frequency and extent of change in VM workloads. For migrating VMs, our proposed algorithm will select appropriate VM(s) to migrate to capable alternative physical machine(s) considering total downtime, overall migration time, maximum number of hosts it will allow to shutdown, etc. We will design our simulation experiments to collect information on total downtime, migration time and power saving with and without hybrid provisioning.

6. Conclusions and Future Work

In this paper we have motivated the need for a hybrid (static/dynamic) VM provisioning approach in clouds, discussed the changes we have made to the Eucalyptus open source cloud system to support hybrid provisioning, and described our initial framework for developing hybrid provisioning heuristics as well as our planned evaluation strategy. In the short-term, we will be implementing our first threshold-driven dynamic provisioning algorithms, verifying their correctness using our testbed system and assessing their basic performance characteristics (e.g. migration overhead) for later use in simulation experiments. We will also be implementing more advanced static provisioning algorithms in our modified version of Eucalyptus to ensure that fair comparisons are possible. We will then explore techniques for making the setting of the framework thresholds dynamic based on global cloud conditions. We then plan to explore two additional extensions to enhance over performance of the framework and algorithms implemented using it: (i) dynamically adjusting the historical periods over which resource use is evaluated, and (ii) working towards global rather than local optimization of migration decisions by using a pre-processing step to identify similarly mis-behaving VMs that should be considered for migration as a group. Finally, we will explore the characteristics of our dynamic migration algorithms that should be used to trigger to re-static provisioning, implement that process and assess its cost and effectiveness.

Acknowledgments

This research was supported in part by the Natural Sciences and Engineering Research Council of Canada under grant number RGPIN 194227-01 and also by the University of Manitoba Graduate Fellowship (UMGF) program.

Journal of Physics: Conference Series **385** (2012) 012004 doi:10.1088/1742-6596/385/1/012004

References

[1] Armbrust M, Fox A, Griffith R, Joseph A D, Katz R, Konwinski A, Lee G, Patterson D, Rabkin A, Stoica I and Zaharia M 2010 *Commun. ACM* **53** 50–58 ISSN 0001-0782 URL `http://doi.acm.org/10.1145/1721654.1721672`

[2] Rimal B, Choi E and Lumb I 2009 *INC, IMS and IDC, 2009. NCM '09. Fifth International Joint Conference on* pp 44–51

[3] Zhang S, Zhang S, Chen X and Huo X 2010 *Future Networks, 2010. ICFN '10. Second International Conference on* pp 93 –97

[4] Meng X, Isci C, Kephart J, Zhang L, Bouillet E and Pendarakis D 2010 *Proceeding of the 7th International Conference on Autonomic Computing* ICAC '10 (New York, NY, USA: ACM) pp 11–20 ISBN 978-1-4503-0074-2 URL `http://doi.acm.org/10.1145/1809049.1809052`

[5] Fito J O, Presa I G and Guitart J 2010 *Parallel, Distributed, and Network-Based Processing, Euromicro Conference on* **0** 111–118 ISSN 1066-6192

[6] Kim H, Kim W and Kim Y 2010 *Grid and Distributed Computing, Control and Automation* (*Communications in Computer and Information Science* vol 121) ed Kim T h, Yau S S, Gervasi O, Kang B H, Stoica A and Dominik (Springer Berlin Heidelberg) pp 84–94 ISBN 978-3-642-17625-8

[7] Rolia J, Cherkasova L, Arlitt M and Andrzejak A 2005 *Proceedings of the 5th International Workshop on Software and Performance* WOSP '05 (New York, NY, USA: ACM) pp 229–237 ISBN 1-59593-087-6

[8] Clark C, Fraser K and Steven H 2005 pp 1–11 URL `http://www.usenix.org/events/nsdi05/tech/clark.html`

[9] Amza C, Chanda A, Cox A, Elnikety S, Gil R, Rajamani K, Zwaenepoel W, Cecchet E and Marguerite J 2002 *Workload Characterization, 2002. WWC-5. 2002 IEEE International Workshop on* pp 3 – 13

[10] Hines M R and Gopalan K 2009 *Proceedings of the 2009 ACM SIGPLAN/SIGOPS International Conference on Virtual Execution Environments* VEE '09 (New York, NY, USA: ACM) pp 51–60 ISBN 978-1-60558-375-4 URL `http://doi.acm.org/10.1145/1508293.1508301`

[11] Inc E S Learn about cloud computing http://open.eucalyptus.com/learn/what-is-eucalyptus accessed on April 25, 2011

[12] Calheiros R N, Ranjan R, Rose C A F D and Buyya R 2009 *CoRR* **abs/0903.2525**

[13] Shi Y, Jiang X and Ye K 2011 *Cluster Computing (CLUSTER), 2011 IEEE International Conference on* pp 595 –599

[14] Foundation A S Olio http://incubator.apache.org/olio/ accessed on April 25, 2011

[15] Bahga A and Madisetti V K 2011 *Journal of Software Engineering and Applications* **4** 396 – 410

[16] EC2 Amazon elastic cloud computing http://aws.amazon.com/ec2/ accessed on March 15, 2011

[17] Voorsluys W, Broberg J, Venugopal S and Buyya R 2009 *Proceedings of the 1st International Conference on Cloud Computing* CloudCom '09 (Berlin, Heidelberg: Springer-Verlag) pp 254–265 ISBN 978-3-642-10664-4

Journal of Physics: Conference Series **385** (2012) 012005 doi:10.1088/1742-6596/385/1/012005

Avoiding the Disk Bottleneck in Deduplicated VM Transfer

Jacob Slack, Samer Al-Kiswany, Matei Ripeanu
The University of British Columbia
Vancouver, BC, Canada

Email: {jacob, samera, matei}@ece.ubc.ca

Abstract. This paper presents an optimization mechanism to increase the performance of cloud services that transfer groups of deduplicated virtual machine (VM) images. This is necessary as the naive data transfer approach for groups of deduplicated VM images is extremely inefficient as it generates highly random disk access pattern. The optimization mechanism presented significantly improves the VM image transfer performance through intelligent buffering, block-level transfer scheduling, and data indexing to increase disk access sequentiality.

1. Introduction

Data deduplication is a commonly adopted optimization in the cloud infrastructure for optimizing the storage [1] and the transfer [2] of virtual machines (VM). Data deduplication efficiently identifies and eliminates similarity across VM image files leading to a reduction in the amount of data that needs to be stored and/or transferred. This reduction in data size is significant: recent studies show that similarity across virtual machines can be as high as 96% [3].

Moving a group of VM images within, or across, data centers is a frequent operation to support application migration, new application deployment, as well as backup and maintenance operations. While deduplication reduces the overall size of a group of VM images it complicates their efficient transfer and re-incarnation at the destination site. The reason is mainly because a single block of deduplicated data can be part of multiple VM images and this introduces coupling between logically independent VM images.

The current naive approach for transferring a group of VMs over the network to a destination site works as follows: the destination site retrieves each block of the deduplicated VM images from the source site and writes it to all the VM images the block is part of at the destination. This approach is embarrassingly inefficient. The reason is that each data block (of typical size of 4-8KB) is often written to multiple VM images generating multiple, seemingly random, disk accesses. This disk access pattern significantly reduces the disk performance, due to the high disk rotation and seek overhead, leading to a low performance for the complete VM transfer process. Experiments show that the performance degradation of this approach is significant leading to transfer rates with throughput in the order of tens of KBps [2] (two orders lower than sequential throughput of commodity disks, and one order of magnitude lower than typical WAN throughput).

This paper presents a mechanism to reduce the randomness in the disk access pattern when transferring groups of deduplicated VM images. This mechanism uses intelligent data buffering and optimized block ordering to increase disk access sequentiality. Analytical analysis of the proposed mechanism shows that the mechanism can bring tangible performance gains compared to the current approaches.

The rest of this paper details the deduplication techniques commonly used in VM image transfer and storage (section 2), surveys work related to the proposed mechanism (section 3), presents the design of the proposed mechanism (section 4), proves through analytical analysis the significant performance gains the proposed mechanism can bring (section 5), and presents experimental evaluation to estimate the performance gains the proposed mechanism can bring and to evaluate the proposed approach effectiveness in eliminating the disk access bottleneck (section 6). We conclude by discussing related issues (section 0).

2. Background

Data deduplication is a commonly used technique for detecting and eliminating similarities across VM images [1, 2, 4]. Deduplication divides the processed files into blocks, computes blocks' identifiers using a collision-resistant hash function such as SHA1 or MD5, and compares these identifiers to detect similar blocks (as equal hashes indicate similar block content).

Two common solutions are adopted for dividing the file into blocks: fixed size blocks, and detecting block boundaries based on file content (a.k.a. variable size blocks). The first solution uses fixed-size blocks: i.e., it divides the file into equal size blocks (e.g. 4KB). The second approach identifies block boundaries (i.e., markers for block start and end) based on file's content. For instance, the Low-Bandwidth File System (LBFS) [5] and JumboStore [6] both detect block boundaries by passing all successive 48 byte 'windows' of the file through a hash function and declaring a block boundary if the last 20 bits of the hash value are all zeros. The advantage of this approach is that, unlike fixed-blocks, the ability to detect block similarities is preserved even in the presence of data insertion and deletion.

The result of deduplication is a set of unique data blocks and a block map for each VM image. Each block map lists the blocks (and their order) that form the image. It is important to note that (in case of high similarity across files) most of the blocks will be part of more than one VM image.

3. Related work

VM Transfer. Recently, a number of VM migration tools use deduplication to reduce the amount of data transferred between the source and destination. Sapuntzakis et al. [4] present a system for virtual machine live migration that includes the VM images. The approach reduces the amount of data transferred using data deduplication between the transferred image and images at the destination. Hirofuchi et al. [7] and Bradford et al. [8] present an optimized VM migration system for migrating live (i.e. running) VM images. Finally, VMFlock [2] presents a deduplication based VM migration system optimized for migration of a group of VM images over WAN.

While the tools surveyed above adopt deduplication for VM migration, they focus on optimizing the system for live migration, without optimizing the disk migration. Our own experience demonstrates that the bottleneck for this use case is in the disk throughput when re-incarnating the VMs at the destination site (surprisingly, the network bandwidth, even in across WAN migrations, or computing hash functions for deduplication, are not bottlenecks).

The mechanism this paper proposes will enable these tools to achieve higher migration throughput.

4. Mechanism Design

The goal of the proposed mechanism is to increase the sequentiality of disk accesses during the VM transfer *operation*. While, the mechanism can be executed at the source or the destination the rest of this section assumes that the mechanism is running at the destination.

Further the proposed mechanism can be used with fixed size blocks as well as content based block boundaries. To simplify the analysis we analyze the mechanism when used with fixed size blocks.

The rest of this section details the mechanism system design.

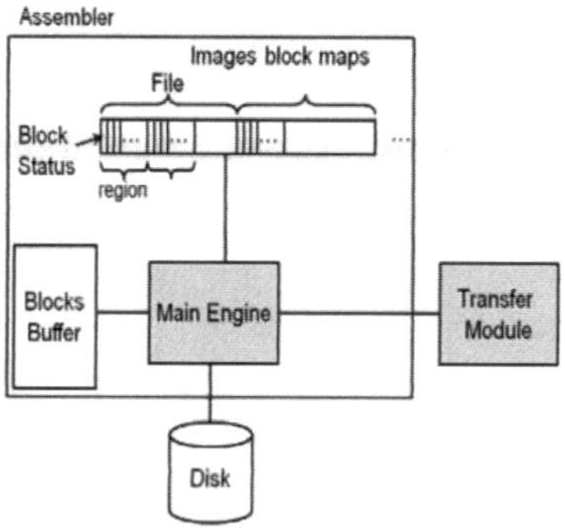

Figure 1. System components. The gray components are active components. The assembler selects blocks to transfer using the Transfer Module, manages the blocks buffer and writes VM images blocks to disk.

A) System Design

The system is composed of two main components (Figure 1): The assembler and the block transfer module.

- *The Transfer Module*: Is the module responsible for implementing the block transfer protocol between source and destination. The module receives the transfer request from the assembler specifying which blocks to transfer, retrieves the blocks from the source, and returns the transferred blocks to the assembler.
- *The Assembler*. Is the central component responsible for assembling the VM images. The main engine in the assembler analyzes the VM blocks to select the next blocks to transfer, manages the block buffer and writes the blocks to the disk. The next subsection details the assembler main engine algorithm. The assembler uses two data structures during the assembly process:
  a. *The Block Buffer*: is an in-memory buffer for data blocks. The Block Buffer is implemented as a hash table accessed using the block identifiers. Each block in the buffer has a count of number of times it shows up across all VM images that are transferred. Every time a block is written to a VM image the counter is decremented. Once the counter reaches zero the block is removed from the buffer. The buffer can hold a maximum of M data blocks. Further, the buffer has a threshold $T<M$ that indicates when the buffer is almost full.
  b. *The Images Block maps*. The system maintains a block map for every image. This shows the order of the blocks in the image. Additionally, each block map is divided into regions composed of b consecutive data blocks. The block map maintains the following information per block: size, identifier (i.e., the block's hash value), number of block occurrences in all images, and status (*at-source*, *buffered* in the block buffer, or *on-disk* at destination).

B) Block maps construction.

System employing deduplication maintain a minimal image block map (contains block hashes and block order) for each image. The Assembler analyzes the minimal block map provided by the source to build a map relating blocks to their list of locations. The analysis includes counting the number of occurrences of each block across VM images using an auxiliary hash table.

C) Assembler Algorithm

The assembler main engine follows the following steps:

1. The assembler retrieves the images minimal block maps from the source. These minimal block maps are produced when the VM images are deduplicated at the source. The assembler analyzes the minimal block maps and creates the extended block maps described in the previous subsection.

2. The assembler retrieves and writes sequentially to the disk the first *k* regions of the first VM image. To maintain sequentiality, any data block in these first *k* regions that is also part of another VM image is buffered in the block buffer.
3. When the number of data blocks in the block buffer exceeds the threshold *T*, the mechanism tries to create more space by writing the buffered blocks to disk and at the same time preserve sequentiality. The assembler analyzes the block maps and selects the region that has the largest number of buffered blocks and starts retrieving and writing blocks for that region. While this decision may free blocks from the data buffer it may also add new blocks as the mechanism retrieves the blocks missing in the region.

   Note that a block maybe part of more than one VM image, consequently, the block is removed from the buffer only when it is written to all the images it is part of.
4. If step 3 succeeds in reducing the number of buffered blocks bellow *T*, the mechanism will continue to retrieve and write sequentially to the current location in the VM image.
5. If step 3 fails to reduce the number of buffered blocks and the buffer becomes full, the assembler selects the region with highest number of blocks in the buffer and writes the blocks from the buffer to the region without bringing any new blocks. While this will not completely transfer the region it will effectively reduce the number of buffered blocks. The mechanism will process as many regions as needed until the number of buffered blocks is less that *T*.

5. Performance Analysis

This section presents an analytical analysis of the expected improvements the proposed mechanism can bring. Table 1 presents the variables used in the proof.

Table 1. Variables used in the performance analysis

F	The total number of VM images to be transferred.
S	The size of a VM image (we assume all images have the same size). The total size of VM images to be migrated is $F \times S$.
b	The number of data blocks per region.
r	The number of regions the data buffer can hold. i.e. the size of the buffer in regions. The number of blocks the buffer can hold is $b \times r$.
c	The number of blocks that always (in every region of every file) appear consecutive. E.g. $c = 2$ means in every region there are two blocks that appear in the same consecutive order in all the other files.
s(d)	The time taken to seek to a new location that is *d* bytes away on the disk. This time equals: disk head seek time + rotation latency. In our analysis we assume that the head seek time is constant. Further we note that rotation latency is directly dependent on the distance the head needs to rotate, *d*.

As the disk seeks overhead is the main contributor to the performance degradation we evaluate the number of seeks required to assemble a set of F images.

To make the analytical analysis tractable and provide an insight on the factors that affect performance, we assume the worst case scenario in which the similarity rate is 100% meaning all the VM images contain the same data blocks (but possibly in a different order), hence every block needs to be written to *F* images at the destination.

Our mechanism's performance is highly impacted by the degree of similarity of regions across files. This is modeled by the parameter c in our model.

Number of seeks in the current approaches:
As each data block is part of the F images under transfer the naïve approach will suffer from F disk seeks for each data block, resulting in total of $(b \times F)$ disk seeks per region regardless of the block pattern; i.e., regardless of the parameter c. Consequently the total seek overhead is: $(b \times F \times s(d_c))$ as d_c denotes the average distance traveled per seek using the current approaches.

Number of seeks generated using the proposed mechanism:
To simplify the analysis of the proposed approach we analyze a simplified version of our mechanism. This provides an upper bound for the number of seeks the original mechanism will require (and thus a lower bound for the performance of our mechanism).

The Simplified Migration Algorithm:
The simplified migration algorithm follows the following steps:

1. Select a region in the file and transfer the blocks in the region sequentially. As all the blocks in the region are part of other F files all the blocks in the region need to be retained in the buffer.
2. When the data buffer is full (i.e. has buffered r regions) the algorithm writes all the buffered blocks to the $(F\text{-}1)$ files they are part of, without bringing any new blocks.

Analysis of the Simplified Migration Algorithm:

- During the first step (point 1) in the simplified algorithm the algorithm will have one disk seek to start writing to a region and then write all the region's blocks sequentially. This leads to one seek per region (or per b blocks).
- After writing r regions the buffer is full. The algorithm will write every block to the $(F\text{-}1)$ files it is part of. Since c of the blocks of every region are consecutive in every file the algorithm will acquire $(F\text{-}1)$ seeks for the consecutive c blocks and $(F\text{-}1)$ for every non-consecutive block. In total the algorithm will acquire $(F\text{-}1) + (b\text{-}c)(F\text{-}1) = (b\text{-}c\text{+}1)(F\text{-}1)$ seeks per region.

In total, the algorithm imposes $(1 + (b\text{-}c\text{+}1)(F\text{-}1))s(d_a)$ seek time to transfer one region. d_a denotes the average distance traveled per seek using the simplified algorithm.

Compared to the naïve approach, the relative performance of the simplified approach in transferring one region is:

$$\frac{\left(1+(b-c+1)(F-1)\right).s(d_a)}{b.F.s(d_c)}$$

This analysis shows that the performance gain is impacted by: the size of the region, the level of region similarity, and the seek distance.

Constant Seek Overhead Analysis
To understand the impact of the region size b and region similarity c on the relative performance we assume that the seek overhead is equal between the two approaches; that is $d_a = d_c$ (we will analyze the effect of the seek distance in the following paragraphs). The relative performance becomes:

$$\frac{1+(b-c+1)(F-1)}{b.F}$$

This formula shows that in the worst case (with $c = 1$, i.e. no two blocks appear in the same region more than once across files), for every seek in the naïve approach the simplified algorithm performs $(F\text{-}1)/F$ seeks. In other terms the simplified approach reduces the number of seeks by $1/F$. This pattern represents the worst scenario for the proposed mechanism as it is less likely to buffer more than one data block in the same region. On the other end of the spectrum, in the best case ($c = b$, i.e. the blocks in a region in a file appear exactly in the same order across files), for every seek in the simplified algorithm the naïve approach performs b seeks; achieving b times performance improvement; this is mainly because the mechanism can efficiently use the buffer to achieve high sequentiality.

Seeks-Distance Impact Analysis
The seek overhead is directly dependent on the distance the disk needs to rotate. The following analyzes the seek distance of the two approaches:

The average seek distance in the current approaches:
In the current approaches every block needs to be written to every file. In the best case the current approaches may impose a seek distance of 0 (in case the current block is the last bock in the current image and the first block in the next image) and a seek distance of $2.S$ in the worst case. The seek distance follows a triangular probability distribution with parameters $a = 0$, $b = 2S$ and $c = S$ with a mean value of S. Consequently, on average the current approaches impose an average seek overhead of $s(S)$.

The average seek distance in the simplified algorithm:
The simplified approach writes the blocks (total of $b\times r$ blocks) in all regions in order (i.e. the blocks are sorted based on their offset). Consequently, assuming a uniform random distribution of blocks across files and offsets the mechanism imposes on average $s(S\times F/b\times r)$ seek overhead.

This analysis shows that the proposed mechanism achieves $F/b\times r$ times shorter seek distance. We note that this is a significant improvement since $r\times b >> F$. The experimental evaluation (section 6) shows that the impact of this improvement on transfer rate is significant.

Conclusion
The proposed approach achieves significant performance gains compared to the naïve approach. This is because it does not only reduce the number of seeks required but it also significantly reduces the seek overhead. For instance for migrating 5 VM images with b set to a reasonable value of 100 4KB blocks and assuming equal seek overhead) the mechanism can achieve up to 100x, and at least 20%, better performance compared to the naïve approach.

Finally, given the nature of VM images, often a group of blocks appear always in the same region across files (i.e., there is high region similarity), since these blocks often belong to the same file in the VM image (same OS kernel, library, or application files). This leads us to believe that the performance gains are considerably higher than the minimum of $1/F$.

6. Experimental Evaluation

This section presents an evaluation of the impact the seek overhead has on the write operation throughput of large VM images and evaluates the effectiveness of the proposed approach in reducing the seek overhead.

We run our evaluation on two machines each with an Intel Xeon E5345 4-core, 2.33-GHz CPU, 4-GB RAM, and a 300GB 7200rpm SATA disks. Each experiment is repeated 10 times and we report average and standard deviation (as error bars).

6.1. Seek Overhead Evaluation

The evaluation writes 10GB files to disk using different block sizes. The evaluation compares three write patterns: sequential, random, and a 'regular' write pattern. For the random pattern the block write order is selected randomly from the possible $n!$ block' orders – where n is the total number of blocks. Using probability analysis, the expected seek distance is 1/3 of the file size (in our evaluation, it is equal to disk distance for storing 3.3GB of data). The regular pattern divides the file into 10,000 segments and writes one block to each segment in a round robin fashion. The goal of this pattern is to measure the impact of seek distance on the disk write throughput. Like the random pattern, this pattern will require a seek for every block write, unlike the random pattern the seek distance is 1/10,000 of the file size (in our evaluation is equal to disk distance for storing 1MB of data).

We note that random pattern represents the pattern generated by the current VM transfer approaches, while the regular pattern represents pattern generated by our mechanism in the worst case (c = 1 in our analysis). Finally, the sequential pattern represents the highest possible performance in our system and approximates the performance of our mechanism in the best case scenario.

We note that our evaluation evaluates the complete disk system including disk and OS caches.

Figure 2 shows the write throughput of the three patterns while varying the block size. The results lead to the following observations:

- Random writes with small block sizes (typical in VM deduplication) achieve embarrassingly low performance (110 KBps with 1KB blocks), up to three orders of magnitudes slower than the sequential write pattern.
- Regular write pattern always performs better than random writes (up to two orders of magnitude better in small blocks) due to the much shorter seek distance.

Summary. Our evaluation shows the low disk performance the current VM transfer mechanism achieves. Further, our evaluation shows that not only increasing disk access sequentiality improves performance, but also minimizes seek distance.

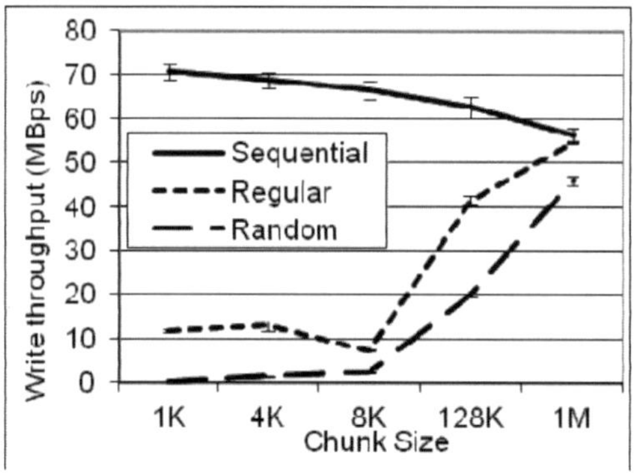

Figure 2. Write throughput of the three patterns. Lines show average throughput of 10 runs, while error bars show standard deviation.

6.2. *Mechanism Evaluation*

To evaluate our approach effectiveness in eliminating the disk random access problem we have built a VM migration tool that implements the approach we propose as well as the naïve migration approach (as a comparison baseline). The prototype uses a single node as a source and a single node as a destination. The destination buffer size was set to 234MB (60,000 4KB blocks).

The evaluation uses two workloads. The main difference between these workloads is in the similarity, at the operating system level. Similarity can vary from using the same operating system installation yet with different applications on top, to using different versions of the same operating system, to using completely different operating systems. Note that all images we use are sparse images; that is, it stores only the blocks of the image disk that are used, not the unused blocks. Consequently, images are often significantly smaller in size than the actual disk size they represent.

The workloads are:

- *Application.* This is a group of three VM images part of OpenCart eCommerce application. The application contains: an image running the OpenCart and an Apache server, one running a MySql server, and an NFS backend storage node. The three images use the same OS distribution: Debian 6.0.4. The total size of all images is 4.4 GB.

- *Different OS*. This is a group of four VM images using different OS distribution. The images contain: Fedora 13 desktop, OpenSUSE desktop, Ubuntu Desktop, Ubuntu server. The total size of all images is 10.6GB. This group of VM images represents the worst case workload for deduplication based migration mechanisms as it contains lowest level of similarity.

To characterize the migration disk access pattern we instrumented the tool to report the offset of the write operation. Analyzing the write offsets enables measuring the ration of sequential write operations.

Our evaluation shows that, when migrating the *application* images, while the naïve approach achieves 15% sequential access by this metric, the proposed approach achieves 99% sequential access. This result highlights the effectiveness of the proposed approach in eliminating the disk bottleneck when migrating a group of VM images. Similarity, for the different OS workload, the naïve approach achieves 9% sequential access, while the proposed approach still achieves 99%.

7. Discussion

1.) *Will this mechanism help if the storage systems at both ends are deduplicated?*

The proposed mechanism optimizes the storage and transfer of a group of VM images in all cases that require assembling the VM images into separate files. This is required not only when transferring a group of VM images from a deduplicated VM repository to a non-deduplicated one, but also when transferring group of VM images across deduplicated VM repositories that do not use the same deduplication technique and configuration, as the incompatibility in deduplication mechanism or configuration can only be addressed by assembling the VM images and deduplicating them again following the destination deduplication mechanism and configuration.

2.) *Is this mechanism tied to any particular VM transfer system?*

The proposed mechanism focuses on intelligent buffering and reordering of VM transferred data blocks. These techniques is not tied to any particular VM transfer system and can be used to optimize systems as simple as single node transfer service to a complex distributed migration system as VMFlock [2].

8. References

[1] D. Reimer, A. Thomas, G. Ammons, T. Mummert, et al. *Opening Black Boxes: Using Semantic Information to Combat Virtual Machine Image Sprawl.* in *ACM Virtual Execution Environments (VEE)*. 2008.

[2] S. Al-Kiswany, D. Subhraveti, P. Sarkar, and M. Ripeanu. *VMFlock: virtual machine co-migration for the cloud.* in *International Symposium on High Performance Distributed Computing (HPDC)*. 2011.

[3] A. Liguori and E. V. Hensbergen. *Experiences with content addressable storage and virtual disks.* in *Workshop on I/O Virtualization (WIOV)*. 2008.

[4] C. P. Sapuntzakis, R. Chandra, B. Pfaff, J. Chow, et al. *Optimizing the migration of virtual computers.* in *Symposium on Operating Systems Design and Implementation (OSDI)*. 2002.

[5] A. Muthitacharoen, B. Chen, and D. Mazieres. *A Low-bandwidth Network File System.* in *Symposium on Operating Systems Principles (SOSP)*. 2001. Banff, Canada.

[6] K. Eshghi, M. Lillibridge, L. Wilcock, G. Belrose, et al. *JumboStore: Providing Efficient Incremental Upload and Versioning for a Utility Rendering Service.* in *USENIX Conference on File and Storage Technologies, FAST*. 2007.

[7] T. Hirofuchi, H. Ogawa, H. Nakada, S. Itoh, et al. *A Live Storage Migration Mechanism over WAN for Relocatable Virtual Machine Services on Clouds.* in *International Symposium on Cluster Computing and the Grid (CCGrid)*. 2009.

[8] R. Bradford, E. Kotsovinos, A. Feldmann, and H. Schioberg. *Live wide-area migration of virtual machines including local persistent state.* in *International conference on Virtual Execution Environments (VEE)*. 2007.

High Performance Computing Symposium 2012 (HPCS2012) IOP Publishing
Journal of Physics: Conference Series **385** (2012) 012006 doi:10.1088/1742-6596/385/1/012006

CUDA accelerated simulation of needle insertions in deformable tissue

Alexandru Patriciu

1280 Main Street West, ITB A308, Hamilton, ON, Canada, L8S 4K1

E-mail: patriciu@mail.ece.mcmaster.ca

Abstract. This paper presents a stiff needle-deformable tissue interaction model. The model uses a mesh-less discretization of continuum; avoiding thus the expensive remeshing required by the finite element models. The proposed model can accommodate both linear and nonlinear material characteristics. The needle-deformable tissue interaction is modeled through fundamental boundaries. The forces applied by the needle on the tissue are divided in tangent forces and constraint forces. The constraint forces are adaptively computed such that the material is properly constrained by the needle. The implementation is accelerated using NVidia CUDA. We present detailed analysis of the execution timing in both serial and parallel case. The proposed needle insertion model was integrated in a custom software that loads DICOM images, generate the deformable model, and can simulate different insertion strategies.

1. Introduction

Needle insertions are used in medical procedures such as biopsies, Radio Frequency Ablations (RFA), or Cryo-ablations. The needle placement accuracy is important for the outcome of the procedure. Needle insertions are usually performed under image guidance which may or may not be real time. During the procedure, the tissue will deform as the needle is inserted; this may cause a shift of the target from the needle trajectory. In order to assess beforehand the accuracy of the needle insertion, it is desirable to simulate the needle insertion.

One of the first studies in needle-deformable tissue interaction was performed by DiMaio [5]. Their approach uses a linear elastic Finite Elemement Model (FEM) to simulate the tissue and the needle. The needle-deformable tissue interaction is modeled using a stick-slip approach. To validate the approach, an experimental system for measuring planar tissue deformation during needle insertions was developed. Goksel [7] extended that model to 3D. The model was also adapted to accommodate arbitrary meshes so that the anatomy can effectively be meshed using third-party algorithms. Nienhuys's [13] approach was also based on FEM and uses quasi-static stick-slip friction for needle-deformable tissue interactions. It uses an iterative relaxation algorithm, and adaptive mesh resolution.

Alterovitz et al. investigated the insertion of flexible needles with beveled tip in deformable tissue [3]. They also proposed a path planning algorithm for beveled needle insertion for prostate. These studies are based on a dynamic FEM formulation and assumed linear models for tissue characteristics. In their recent work [2], they consider uncertainty and the planner computes optimal actions to maximize the probability that the needle will reach the desired target.

Misra et al. [12] proposed a new model of flexible needle-deformable tissue interaction. Their work extends the non-holonomic kinematic model proposed by Webster et. al. [14] by taking into consideration the microscopic interaction between the bevel tip of the needle. Their results show that the models match the macroscopic needle insertions observations but discrepancies were especially large for large bevel angles. Duriez et al. [6] used complementarity constraints for the simulation of the insertion of flexible needles into soft tissue. Dehghan et al. [4] presents a needle path planning method for the insertion of rigid needles into deformable tissue. The optimization method is based on iterative simulations performed using a tissue finite element model. The experiment showed substantial decrease in the targeting error.

Within the computational mechanics community a strong research interest is devoted to the mesh-less methods that emerged in the last two decades. A comprehensive review of the mesh-less approach to continuum mechanics models is presented by Li and Liu [10]. Horton et al. [8] developed a meshless method for simulating soft organ deformation and simulated indentation of a swine brain and compared the results to experimental data. Li et al. [9] presented an adaptive meshless method (MLM) for solving deformable contact problems and validated it against analytical and numerical solutions.

In this paper we consider the interaction between stiff needles and deformable tissues. This is relevant for applications like breast biopsy. The deformable tissue is modeled using a mesh-less method (RKPM) and the needle-deformable tissue interaction is modeled using fundamental boundaries. The forces are generated either by the friction and cutting or by the needle constraints. In contrast with the approach followed by Duriez [6] the constraints are not solved explicitly but they are instead dynamically adjusted during the insertion.

The paper is organized as follows Section 2 presents the needle-deformable tissue interaction model; deformable tissue model and details regarding the parallel implementation on the Graphical Processing Units(GPU); Section 3 provides simulation results. The paper concludes with some observations and future work.

2. Stiff needle-deformable tissue interaction model

Several needle-deformable tissue interaction models have been proposed by researchers, a comprehensive survey of these models can be found in the paper by Abolhassani et al. [1] and Misra et al. [11].

In the proposed approach we model the needle-deformable tissue interaction as a fundamental boundary. The needle insertion action is performed by applying an insertion force on the needle. The interaction between the needle and the tissue comprises three types of forces - friction forces, cutting force, and constraint forces. The friction force is aligned with the needle direction and is distributed along the needle. The cutting force is aligned with the needle trajectory and it is applied by the tip of the needle. The sum of friction and the cutting forces gives the insertion force. The constraint force is orthogonal to the needle trajectory and ensures that the tissue in the neighborhood of the needle doesn't move across the needle. From a physical perspective in order to simulate the tissue deformation during the needle insertion we need to model the deformable object subject to the interaction forces.

Figure 1 shows an object in the deformed configuration with the needle inserted in it. The needle trajectory comprises a set of points $pn_i; i = 1 \dots NP$ represented in undeformed body coordinates. Each needle point pn_i has an associated coordinate system ξ^i defined such that ξ^i_0 gives the direction of the needle and ξ^i_1 and ξ^i_2 are orthogonal to the needle direction. In this approach the force applied by the needle to the deformable object through friction and cutting is always aligned with axes ξ^i_0. Similarly, the constraint forces will be aligned with axes ξ^i_1 and ξ^i_2.

For the case of stiff needles all needle model points should lie on the needle axis. In Figure 1 δ^i_1 and δ^i_2 are the deviations of point pn_i from the needle axis along directions ξ^i_1 and respectively ξ^i_2. Similarly, f^i_0 is the axial force applied by the needle, f^i_1 and f^i_2 are the constraint forces. The sum of all axial forces is equal with the insertion force. In order to have a proper configuration,

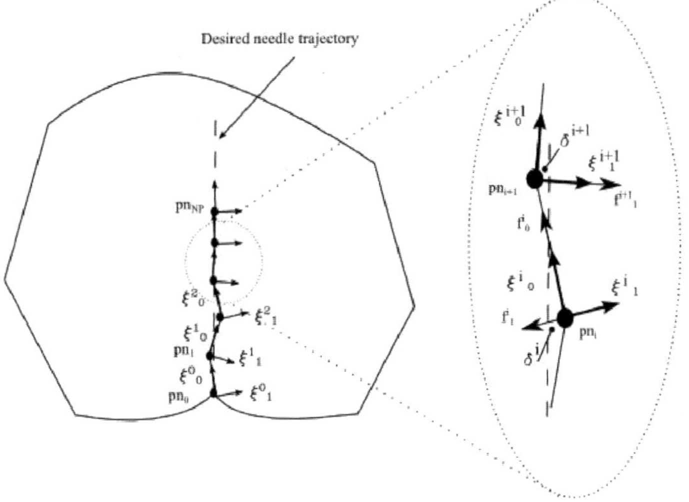

Figure 1: 2D deformable object and needle points. $(pn_i)_{i=1...NP}$ are the needle points represented in undeformed body coordinates; $(\xi^i)_{i=1...NP}$ represent the needle coordinate systems, $(\xi_0^i)_{i=1...NP}$ are aligned with the needle axis, $(\xi_1^i)_{i=1...NP}$ are orthogonal to the needle axis; $(\delta_i)_{i=1...NP}$ represent the distances between $pn_i + u(pn_i)$ and the needle axis; $(f_1^i)_{i=1...NP}$ are the constraint forces

the constraint forces f_1^i and f_2^i have to be adjusted such that $\delta_1^i \to 0$ and $\delta_1^i \to 0$. The simulation starts with zero insertion force, the needle has only one segment inside the tissue. The insertion force is incremented and distributed as friction and cutting force - f_0^i. The deformation is computed and if the needle points deviate from the desired trajectory the constraint forces are updated. This approach has the advantage that requires only a general purpose solver for a deformable object - for example a preconditioned conjugate gradient solver. The object structure will not change during the simulation, only the interaction forces will change. This is a desirable property when a parallel implementation is sought.

2.1. Adaptive Constraints Adjustment

Let's assume that the lumped stiffness of the point pn_i along direction ξ_j^i is k_j^i; if we increment the force along the direction ξ_j^i with δf_j^i this should result in a incremental displacement $\delta f_j^i / k_j^i$. The friction coefficient for the segment (pn_i, pn_{i-1}) is μ_i. Furthermore, let's assume that the function $Solve(pn, f, u)$ implements a solver of the deformable object that applies forces $\mathbf{f^i}$ along the boundary described by the points pn_i and returns the displacements u_i of points pn_i. Then, an iterative algorithm that simulates the needle insertion is described by Algorithm 1.

In this approach the constraints forces propagate from one simulation iteration to the other; therefore only small adjustments are required. The compliances are also propagated and adjusted dynamically during the simulation. The compliance adjustment is performed using the following scheme

Algorithm 1 Needle insertion model simulation workflow

Initialize vector pn with the entry point and a second point in the direction of insertion; $NP = 2$
Initialize constraint forces $f_j^i \leftarrow 0$
Initialize compliances $k_j^i \leftarrow K_0 = 0.1$
$f_{Insertion} \leftarrow 0$
while $f_{Insertion} < f_{Max}$ **do**
  Distribute $f_{Insertion}$ over the needle segments such that $\sum f_0^i = f_{Insertion}$.
  repeat
    $Solve(pn, f, u)$
    Compute deviations from needle trajectory δ_j^i; $i = 1 \ldots NP$; $j = 1, 2$
    Adjust constraint forces $f_j^i \leftarrow f_j^i + \delta_j^i * k_j^i$; $i = 1 \ldots NP$; $j = 1, 2$
    Adjust lumped stiffness coefficients k_j^i; $i = 1 \ldots NP$; $j = 1, 2$
  until $RMS(\left(\delta_1^1, \delta_2^1, \ldots, \delta^N P_1, \delta^N P_2\right)) < \epsilon$
  if $f_0^{NP} >$ CuttingThreshold **then**
    Add new point to the **pn** vector. $NP \leftarrow NP + 1$
  end if
  $f_{Insertion} \leftarrow f_{Insertion} + f_{Increment}$
end while

$$k_j^i = 0.5\left(k_j^i + \frac{f_j^i - fOld_j^i}{\delta_j^i - \delta Old_j^i}\right); i = 1 \ldots NP; j = 1, 2 \tag{1}$$

where $fOld_j^i$ is the previous force δOld_j^i is the previous alignment error. The proposed approach works with any computational deformable model that can accommodate changing fundamental boundaries described in material coordinates.

2.2. Deformable Object Model

Our implementation uses a quasi-static mesh-less deformable model. Both nonlinear and linear versions were implemented. Throughout the paper the following conventions will be followed. The initial (un-deformed) coordinate system is represented by uppercase X and the deformed configuration is represented by lower case x. The region occupied by the body in the initial configuration is Ω_X and it has a boundary Γ_X; $u(X, t) = \phi(X, t) - X$ is the material displacement. The non-linear elasto-static equations were discretized using the reproducing particles kernel method (RKPM). In a RKPM the deformation is defined using global interpolants [10]. The interpolants can be seen as smoothed dirac functions centered in a certain body coordinate. If there are N_P particles distributed over the body. Each particle has an associated shape function $\mathcal{N}_I(X) : \Omega_X \to \mathbf{R}$. Then, the particle displacements can be expressed as $u_i(X) = \sum_{I=1}^{N_P} \mathcal{N}_I(X) d_{iI}$. The incremental discretized equation has the form

$$K_T(d_k)\Delta d = f_{ext} - f_{int}(d_k) \tag{2}$$

where $K_T(d_k)$ is the tangent stiffness matrix, $f_{int}(d_k)$ is the vector of internal elastic forces in the current configuration, and f_{ext} is the force applied through the fundamental boundary. The vector of external forces f_{ext} is composed of the axial and constraint forces applied by the needle. The displacement of the needle points is returned by the solver as $u(pn_i) = \sum_{I=1}^{N_P} \mathcal{N}_I(pn_i) d_I; i = 1 \ldots NP$.

2.3. Parallel implementation

The parallel implementation uses a NVidia GTX-480 GPU. This GPU can be seen as very power-full parallel processing cards connected through PCIe-x16 interface to the host computer. One board has a set of multiprocessors connected to a shared memory of approximately 1.5GB. Each multiprocessor has 32 processing cores which execute the same code and can access a fast shared memory. NVidia CUDA consists of a set of C extensions that enables the programmer to launch threads on the GPU. CUDA model requires the programmer to divide the work in threads that can be executed with minimum inter-thread synchronization. Ideally, all threads should be independent. Deformable object models solvers are suitable for this type of implementations. Algorithm 2 describes the work-flow of the parallel solver.

Algorithm 2 Overall GPU solver workflow

Copy d from CPU to GPU
Copy f_{ext} from CPU to GPU
repeat
  Start GPU threads for computing strains and stresses over the body; One thread will compute the stress in one particle location. The result is deposited in GPU memory.
  Wait for the threads to finish
  Start GPU threads for computing K_T. One thread will compute the K_T block corresponding to a pair of overlapping particles. The result is deposited in GPU memory.
  Start GPU threads for computing f_{int}. One thread will compute the internal force applied to a particle. The result is deposited in GPU memory.
  Wait for the threads to finish
  Solve equation 2 for $\Delta dGPU$ using a PCG algorithm. This algorithm is also implemented on GPU.
  Perform $dGPU_{k+1} \leftarrow dGPU_k + \Delta dGPU$ on the GPU. Compute $\|\Delta dGPU\|$.
until $\|\Delta d\| < \epsilon$
Copy d from GPU to CPU

The parallel algorithm performs the same high level iterations as the serial algorithm. The speed-up is achieved by implementing in parallel the tangent stiffness matrix and internal forces vector computations as well as the preconditioned conjugate gradient solver. We used a data parallelization technique to convert the serial algorithm into a parallel one. Essentially, one thread will compute the block of the K matrix corresponding to two interacting particle. Similarly, for the f_{int} vector, one thread will compute the internal force corresponding to one particle. The K_T matrix is stored as a sparse matrix in ELL format, therefore the sparse matrix vector multiplication can be performed efficiently. This approach ensured that the data transfer between the host computer memory and GPU memory was minimized. Essentially, for each nonlinear solver iteration we have to copy from the GPU memory to the device memory the vector of particle positions which has the size $N_P * n_{sd} * sizeof(float)$ bytes; where n_{sd} is the number of space dimensions.

3. Results

The needle insertion in 2D models was implemented together with a GUI. The interface allows the user to load a MR image or slice. The user defines the fixed boundaries of the object then the object is discretized using a regular grid. Material properties are assigned based on the MR image contrast. For a particle, we first compute the average value of 9 surrounding pixels. If the value is higher than a threshold f_T then the material properties for that particle are assigned to fat, if they are below a certain threshold there is no material in that area. If they are in-between; they are considered stiff nodules.

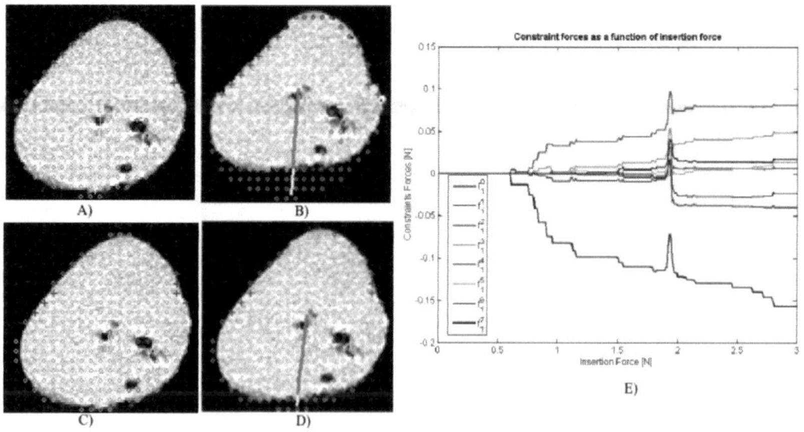

Figure 2: Needle insertion simulation A) before the insertion; B) nonlinear solver after the insertion; C) before the insertion; D) linear solver after the insertion; E) Constraint forces as a function of the insertion force; Red crosses are fixed boundaries; Green circles represent particles with fat material properties; Magenta circles represent particles with stiff material properties

	Serial implementation	GPU Implementation
Update Stresses; Assemble K_T and f_{int}	1698ms	2ms
Mean PCG algorithm time	490ms	5.3ms
Mean Total Solver Time	6s	32ms

Table 1: Timing of different algorithm functions

Then the user has the opportunity to manually adjust the material properties for desired locations. Then, he can select the needle entry site and the insertion direction. The needle insertion simulation is executed while the MR image is warped to display the deformable object.

The material properties were assigned as follows: Lame coefficients for fat: $\lambda = 1.6$, $\mu = 0.03$; Lame coefficients for stiff insertions: $\lambda = 16$, $\mu = 0.3$; Friction coefficient fat $= 0.05N/mm$; Friction coefficient fat $= 0.1N/mm$; Cutting force fat $= 0.5N$; cutting force stiff insertions $= 1N$.

Several simulations were performed. The maximum insertion force was $3N$ and the insertion step force step $0.01N$. The first set of simulations were performed using a nonlinear solver. Figure 2A presents the non-deformed object with the desired needle insertion direction. Figure 2B presents the object state after the needle insertion. The RMS of needle deviations at the end of the simulation was 0.08mm. The maximum displacement over the entire model was $13mm$.

Table 1 presents the timing of each function of the iterative solver. The data shows that the biggest gain is in the K_T matrix assembly time. The data shows that nonlinear insertions simulation can be performed with update rates larger than 20Hz.

From Table 1 it can also be noticed that a linear solver can sustain update rates of almost 200Hz. This is due to the fact that the linear solver has to execute only one PCG solver per iteration. However, Figures 2C and D show that a linear solver underestimate the deformation; this is in agreement with previously reported results. Therefore, in order to achieve accurate simulations a nonlinear model is required.

The constraints force evolution as a function of needle insertion force is presented in Figure 2E. The plot shows a smooth constraint force adaptation during the insertion. The only perturbation occurred when the needle penetrated a stiff insertion. Even in this case, the algorithm recovered and the final deviation from the straight trajectory was under $0.1mm$.

4. Conclusions

The paper presents a needle-deformable tissue interaction model using constraint forces. The advantage of the proposed model is that it requires only a general purpose deformable object solver. This provides avenues for the parallelization of the implementation.

The paper presents also numerical results of needle insertions in 2D deformable object. The results include serial as well as GPU accelerated solvers for both nonlinear and linear elasticity. The numerical results illustrate that in order to achieve accurate simulations a nonlinear deformable model has to be used. This is in agreement with previously reported results.

The work will be extended to 3D models as well as deformable needles. More work is required to compare the simulation results against real life needle insertion experiments.

Acknowledgement: The work was partially supported by Natural Sciences and Engineering Research Council and Canadian Foundation for Innovation.

5. References

[1] N. Abolhassani, R. Patel, and M. Moallem. Needle insertion into soft tissue: A survey. *Medical Engineering & Physics*, 29(4):413–431, May 2007.

[2] R. Alterovitz, M. Branicky, and K. Goldberg. Motion planning under uncertainty for image-guided medical needle steering. *The International Journal of Robotics Research*, 27(11-12):1361–1374, 2008.

[3] R. Alterovitz, K. Goldberg, J. Pouliot, R. Taschereau, and I-Chow Hsu. Needle insertion and radioactive seed implantation in human tissues: simulation and sensitivity analysis. In *Robotics and Automation, 2003. Proceedings. ICRA '03. IEEE International Conference on*, pages 1793 – 1799, 2003.

[4] E. Dehghan and S. Salcudean. Needle insertion parameter optimization for brachytherapy. *IEEE TRANSACTIONS ON ROBOTICS*, 25:303 – 315, 2009.

[5] S.P. DiMaio and Salcudean S.E. Needle insertion modelling for the interactive simulation of percutaneous procedures. *Lecture Notes in Computer Science*, 2489:253–260, 2002.

[6] Christian Duriez, Christophe Gubert, Maud Marchal, Stephane Cotin, and Laurent Grisoni. Interactive simulation of flexible needle insertions based on constraint models. In *Medical Image Computing and Computer-Assisted Intervention MICCAI 2009*, volume 5762 of *Lecture Notes in Computer Science*, pages 291–299, 2009.

[7] O. Goksel, S. Salcudean, and S. DiMaio. 3d needle-tissue interaction simulation for prostate brachytherapy. *Lecture Notes in Computer Science*, 3749:827–834, 2005.

[8] A. Horton, A. Wittek, and K. Miller. Subject-specific biomechanical simulation of brain indentation using a meshless method. *Medical Image Computing and Computer-Assisted Intervention - MICCAI 2007, Pt 1, Proceedings*, 4791:541–548, 2007.

[9] Q. Li and K. Lee. An adaptive meshless method for modeling large mechanical deformation and contacts. In *Robotics and Automation, 2007 IEEE International Conference on*, pages 1207 – 1212, 2007.

[10] Shaofan Li and Wing Kam Liu. Meshfree and particle methods and their applications. *Applied Mechanics Reviews*, 55(1):1–34, January 2002.

[11] S. Misra, K. Ramesh, and A. Okamura. Modeling of tool-tissue interactions for computer-based surgical simulation: A literature review. *Presence*, 17(5):463491, 2008.

[12] S Misra, K B Reed, B W Schafer, K T Ramesh, and A M Okamura. Mechanics of flexible needles robotically steered through soft tissue. *The International Journal of Robotic Research*, online, 2010.

[13] H. Nienhuys and van der Stappen A. A computational technique for interactive needle insertions in 3d nonlinear material. In *IEEE International Conference on Robotics and Automation, 2004*, 2004.

[14] R. J. Webster, J. S. Kim, N. J. Cowan, G. S. Chirikjian, and A. M. Okamura. Nonholonomic modeling of needle steering. *International Journal Of Robotics Research*, 25(5-6):509–525, May 2006.

Parallel algorithm for fractal model of creation of soil structure using particles-size distributions

V Roubtsova, B Morin, M Chekired

Institut de recherche d'Hydro-Quebec, 1740 boul. Lionel-Boulet, Varennes (Quebec) J3X 1S1

e-mail: roubtsova.varvara@ireq.ca

Abstract. To simulate the soil in 2-D we use a fractal scaling of the soil particle-size distribution. The sequential algorithm of this model is based on an 11 levels deep nested loop in which the inner limits are dependent on the previous iteration. In this article we describe the parallel algorithm for this type of problem which avoids the dependent recursion by using a flattened generator.

1. Introduction

Each year, many disasters happen in the word because of interaction of water with soils particles. Dam failures, blurring of road coating, destruction of houses caused by the liquefaction of foundations, all are the sources of not only of physical damages but also of the loss of human lives. The cause of such catastrophes at a macro level which can be found at the micro level is very difficult to observe.

The main objective of our project is to provide simulation tools for both engineers and researchers to freely explore an area of research that would otherwise be very difficult to achieve. We are thus developing a virtual laboratory for geotechnical research. The advantages of this laboratory are (i) access to parameters for which measurement in real experiments are impossible, (ii) the possibility to make experiments which would otherwise be difficult or impossible to recreate in laboratories (either by being too expensive, too long, too difficult to measure or too difficult to recreate), (iii) errors made by engineers or researchers can be corrected at only the cost time.

The Discrete Element Method (DEM) was used to model the soil. This Lagrangian method takes into account all sorts of interaction forces between the particles. We can watch an individual particle: we know its velocity, its location and the forces acting on it at each moment. The fluid motion is based on the Navier-Stokes equations using Marker-and-Cells (MAC) method. This Eulerian approach gives access to information about the pressure and real velocity at each point in a pore-channel.

The visualization plays an important role in the validation of numerical algorithms. Using Alioscopy technology, we were able to project eight different perspectives on a screen covered by a lenticular array, resulting is seven simultaneously observable angles of vision. This technology allows multiple observers to interact with a simulation, and permits more complex systems to be viewed. The OpenGL (Open Graphics Library) API (Application Programming Interface) was used to compute both the flat projection, as well as the different views required by the Alioscopy technology in a portable manner. GLSL (Graphics Library Shading Language) shaders were used to combine the views into a single image projected on the screen, which is then redecomposed into individual images by the lenticular lens.

One of the possible applications for this virtual laboratory is the computation of soil permeability. Soil permeability is one of the most important parameter and is used to preview the soil behavior under water flow. It is dependent only on the network of the pore channels. Our laboratory can helps to determine this parameter by using only the particles-size distribution. The random distribution of particles, with both diameter and quantity corresponding to the particle-size curve, guarantees a good fit with actual soil in 3-D. We have the opposite situation in 2-D, where we need to create an artificial model of pore channels network. There are a lot of different models of pores channels [1, 2, 3, 4], we chose the pore-soil fractal approach [4]. This model offers a wider scope for modeling of soil.

2. Fractal model

The fundamentals of fractal geometry are described by Mandelbrot [5]. Fractals are very popular in soil research as a means of characterizing various properties of porous media. For examples: the fractal pore-size distribution was presented by Friesen and Mikula [6], the Sierpinski carpet was used to model the soil water retaining curve by Soto and Vilar [7], the pore-solid interface was modeled by fractals by Toledo et al. [8], the thermal conductivity of soils was calculated by Lehmann et al. [9] . More information about the different applications of the fractal in soil science is presented in [10].

Initially, soil is represented as a random matrix composed of both particles and pores. In the 2-D model, we need to distribute particles within the matrix in such a way as to obtain the same porosity and conductivity as in an actual 3-D soil.

A detailed study of the Pore Solid Fractal (PSF) model has been conducted by Perrier et al. [4]. The generation of a PSF starts with the initiator, which consists of a region that forms a square surface of length L in a 2-D configuration. This region is divided into R^2 subregions with length L/R. The subregions represent the pores, the particles, and the smaller subregions that will be divided in the next iteration step.

This structure can be described by a generator that includes the scaling factor R, the number of pores P, and the number of particles (solids) S. The number of unspecified elements F (fractals) is equal to R3 – P – S.

This model was extended by combining three fractal processes (generators) in series by Lehmann et al. [9] . Each generator was applied for a certain number of iterations. The resulting three-generator-pore-solid-fractal (3G-PSF) approach accounts for a wide range of particle-size distribution.

We used porosity as the marker of good agreement with natural pore channel network because this parameter represent a more natural fit that the pore-size distribution. For example there is a big difference between definition of pore-size distribution in [11] and [12].

To obtain a more realistic model, we replaced the solid square subregions by subregions containing a circular particle surrounded by an empty area. After some iterations, the porosity is found as follows:

$$\phi_{n1+n2+n3} = \frac{P_1 + (1-0.25\times\pi)S_1}{P_1+S_1}\left(1-\left(\frac{F_1}{R_1^2}\right)^{n1}\right) + \left(\frac{F_1}{R_1^2}\right)^{n1}\frac{P_2+(1-0.25\times\pi)S_2}{P_2+S_2}\left(1-\left(\frac{F_2}{R_2^2}\right)^{n2}\right)$$
$$+\left(\frac{F_1}{R_1^2}\right)^{n1}\left(\frac{F_2}{R_2^2}\right)^{n2}\frac{P_3+(1-0.25\times\pi)S_3}{P_3+S_3}\left(1-\left(\frac{F_3}{R_3^2}\right)^{n3-1}\right)+\left(\frac{F_3}{R_3^2}\right)^{n3-1}\left(\frac{P_3+F_3+(1-0.25\times\pi)S_3}{R_3^2}\right)$$

the indices 1, 2, and 3 correspond to the first, second, and third generators respectively end n is the number of iterations. All unspecified elements are considered to be pores for the last iteration.

In the cases of low porosity we don't have the pore subregions but we have the pore channels that assure the passing of water through soil matrix. The last one is very important for compacting permeable soils.

Finally, the best approximation of the particle size distribution curves and the porosity are determined by the parameters of the generators R_1, P_1, S_1, n_1, R_2, P_2, S_2, n_2, R_3, P_3, S_3, n_3.

Looking at figure1 we can see the steps of the development of a fractal porous medium that corresponds to a typical particle-size distribution. The first generator ($R_1 = 3$, $S_1 = 3$, $P_1 = 4$) was used for the first three steps and it approximate the section of the curve for coarse particles. Using the second generator ($R_2 = 4$, $S_2 = 5$, $P_2 = 7$) for the next two steps, we approximate the middle section of the curve. Finally we used the third generator ($R_3 = 3$, $S_3 = 2$, $P_3 = 2$) to approximate the last part of the curve, representing fine particles.

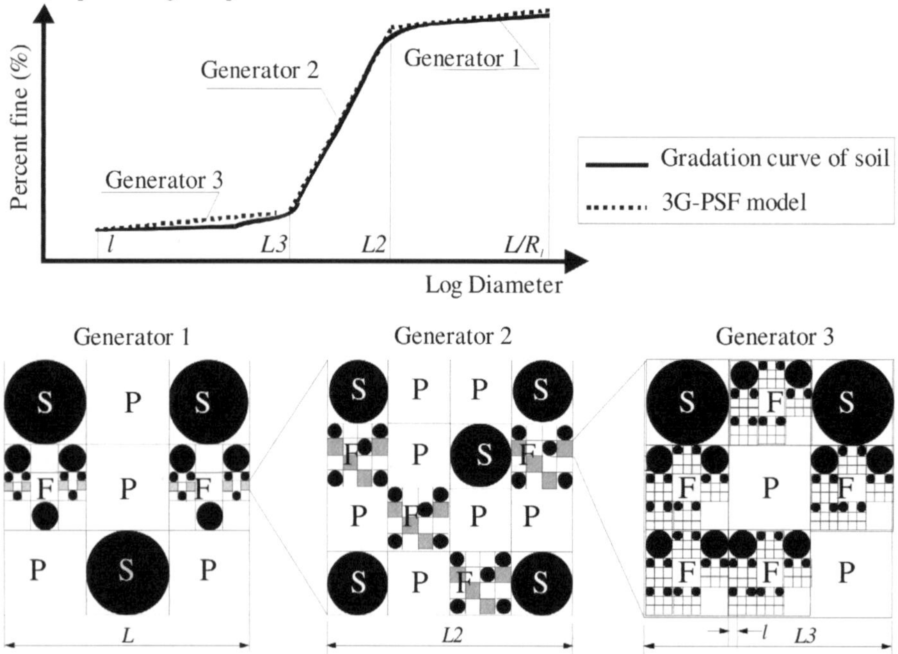

Figure 1. Construction of a 3G-PSF with round particles.
For the first generator: $R_1 = 3$, $S_1 = 3$, $P_1 = 4$, *and* $n_1 = 3$; *for the second generator:* $R_2 = 4$, $S_2 = 5$, $P_2 = 7$, *and* $n_2 = 2$; *and for the third generator:* $R_3 = 3$, $S_3 = 2$, $P_3 = 2$, *and* $n_3 = 3$.

The location of particles into pattern was determined by the following conditions: (i) the fractal structure must have more possible channels from left to right, (ii) the channels mustn't be straight. It should be noted that different location patterns exist for the same parameters. Figure 2 shows examples of soil pattern for different maximal scaling factors.

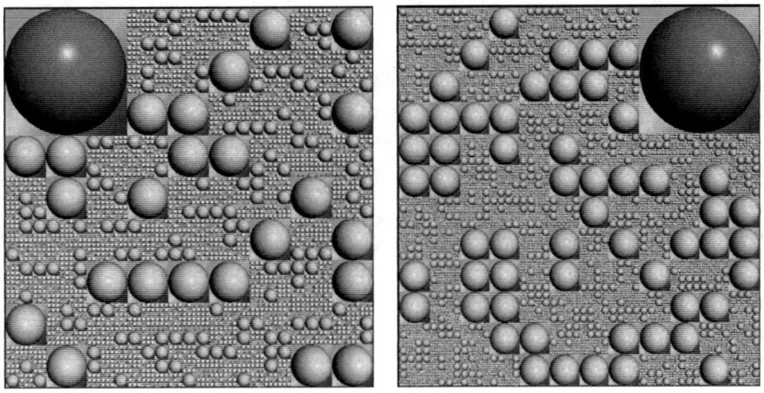

Figure 2. Examples of a soil pattern for $R_{max} = 3$ and $R_{max} = 4$.

The results obtained by this model have been presented by Chekired and Roubtsova [13]. A summary of key achievements are listed below.

A comparison of the particle size distribution of various soil samples with those obtained by fractal modeling is shown in Figure 3.

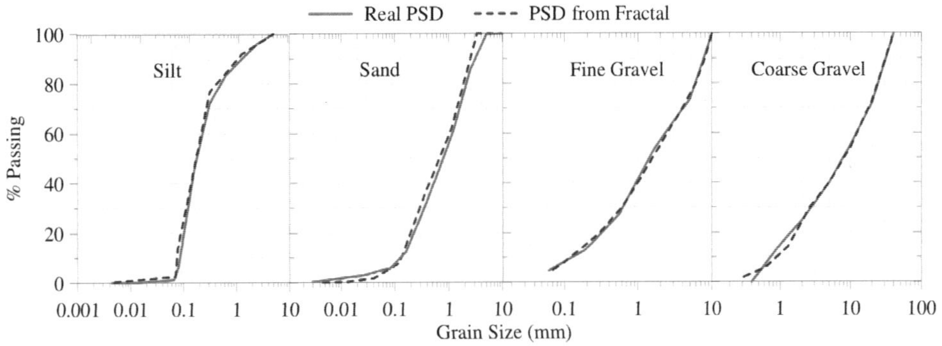

Figure 3. Comparison of PSD for various soil samples with those obtained from the fractal model.

The water flow through a pore channel is described by the Navier-Stokes equations for viscous-incompressible liquids, and is expressed by:

$$\rho\left[\frac{\partial \vec{v}}{\partial t} + (\vec{v}\nabla)\vec{v}\right] = -\nabla p + \eta\nabla^2\vec{v} + \vec{f}$$

$$\nabla\vec{v} = 0$$

where ρ is the density of the liquid (kg/m^3), t is the time (s), $\vec{v}$ is the flow velocity (m/s), p is the water pressure (Pa), η is the dynamic viscosity (Pa·s), and f is the force of gravity (N). Water flows through the pattern from left to right by applying a pressure differential as boundary conditions. Hydrodynamic flows through four different soil samples generated by the fractal model are presented in Figure 4.

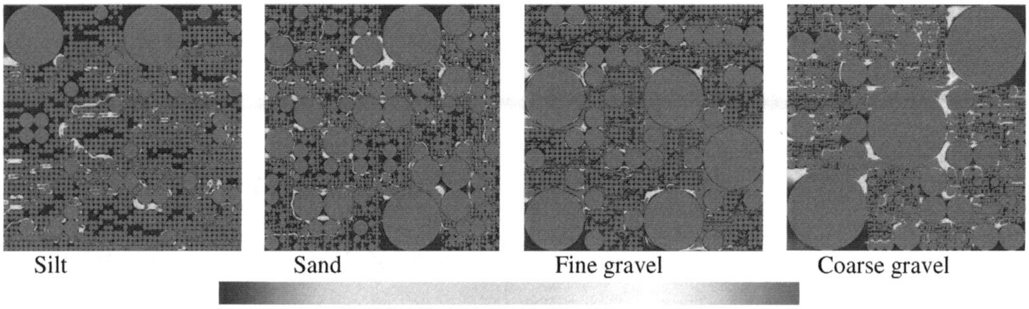

Figure 4. Hydrodynamic flow through generated soils

The coefficient of permeability (m/s) for this soil sample is (Scheidegger [14])

$$K = -\frac{\langle v \rangle s \rho g}{\Delta p}$$

where $\langle v \rangle$ is the average velocity in the soil (m/s), s is the length of the sample in the direction of flow (m), and Δp is the pressure difference at the boundary (Pa).

A comparison of porosity and permeability obtained from the fractal model and from empirical formula for different soils is presented in Tables 1 and 2 respectively.

Table 1. Comparison of porosity

	Silt	Sand	Fine Gravel	Coarse Gravel
Empirical	0.406	0.303	0.261	0.268
Fractal	0.405	0.314	0.265	0.279

Table 2. Comparison of coefficient of permeability (m/s)

		Silt	Sand	Fine Gravel	Coarse Gravel
Empirical		$1.07 \cdot 10^{-04}$	$2.03 \cdot 10^{-04}$	$2.17 \cdot 10^{-04}$	$5.09 \cdot 10^{-03}$
Fractal	Min	$1.61 \cdot 10^{-04}$	$2.18 \cdot 10^{-04}$	$8.61 \cdot 10^{-05}$	$9.68 \cdot 10^{-04}$
	Max	$1.94 \cdot 10^{-04}$	$2.65 \cdot 10^{-04}$	$3.97 \cdot 10^{-04}$	$3.44 \cdot 10^{-03}$

The empirical porosity of the soil was calculated using $\phi = 0.255\left(1 + 0.83^{U}\right)$ [15], where $U = d_{60}/d_{10}$ is the coefficient of grain uniformity and where d_{60} and d_{10} represent the grain diameter in mm at which 60% and 10% grains are finer respectively. For all type of soil, six different pattern where created and the different values for the coefficients of permeability were obtained. We present the both minimum and maximum values for the coefficient of permeability computed during simulations for various particle positions. It is not proven that we take in account all possible placements of particles.

The empirical coefficient of permeability was calculated using the Pavchich formula [11] expressed as follows:

$$K = \frac{g}{\upsilon} 4 \cdot 10^{-3} B \sqrt[3]{U} \frac{\phi^3}{(1-\phi)^2} d_{17}^2$$

where g is the acceleration due to gravity, ν is the kinematic viscosity of water (m^2/s), B is equal to 1 for sand and gravel and 0.35-0.4 for pebbles, and d_{17} represents the grain diameter in mm for which 17% of the sample are fine then. This formula has been found to be satisfactory for a wide range of soils.

These results indicate that the proposed 2-D model can virtually generate different kinds of soils successfully using only grain size distribution curves, and can estimate soil characteristics, such as the coefficient of permeability.

3. Parallel algorithm to create the soil pattern

As notice early we need to check all possible combinations of 12 variables (R_1, P_1, S_1, n_1, R_2, P_2, S_2, n_2, R_3, P_3, S_3, n_3) to create the soil pattern with the best approximation for particle-size curve and porosity. The user can select only the maximum scaling factor R_{max}. The numbers of possible combinations for other parameters are determined by this factor and the particle-size distribution.

The pseudo code for sequential algorithm is:

```
for R1 = 2 to Rmax
 for n1 = 0 to n1max  (n1max =f(R1))
  for R2 = 2 to Rmax
   for n2 = 0 to n2max  (n2max =f(n1max, R1, R2))
    for S1 = 0 to R1^2-1
     for P1 = 0 to R1^2 - S1
      for S2 = 0 to R2^2-1
       for P2 = 0 to R2^2 - S2 - 1
        for R3 = 2 to Rmax
         for S3 = 0 to R3^2-1 do
          for P3 = 0 to R3^2- S3 -1
          {
            Computation distribution and porosity
          }
```

Where:

$$n_{1\max} = -\frac{\log\left(\frac{d_{\min}}{L_1}\right)}{\log(R_1)} + 1$$

$L_1 = d_{\max} R_1$ is the length of matrix;

$$n_{2\max} = \left(-\frac{\log\left(\frac{d_{\min}}{L_2}\right)}{\log(R_2)} + 1\right) \cdot \frac{\left(R_1^2 + R_1^4\right)}{4}$$

$L_2 = \frac{L_1}{R_1^{n_1}}$ is the length of matrix for second generator;

d_{max} is the maximal size of soil particle;

d_{min} is the minimal size of soil particle.

The twelfth variable n_3 only has a single value for each combination of the other variables. It is thus the dependent parameter.

We can note that with R_{max} only equal to three, we obtain more than ten millions possible combinations. It is thus important to decrease the time to parse the results.

Using this algorithm, the computation of porosity and particle-size distribution are independent for each combination of fractal parameters while using the same instructions for all combinations. This type of problem is ideal candidate for parallelization. The main difficulty in parallelizing this algorithm lies in the fact that the inner limits are dependent on the previous iteration values. Our algorithm avoids this recursive dependency by using a flattened generator.

Firstly, we need to know the total number of treads. It is:

$$k = k1 \cdot k2 \cdot k3$$

where

$k1 = \sum_{R=2}^{R\max} n_{1\max}(R_1)$ number of iterations for loops 1-3 in the sequential algorithm;

$k_2 = \sum_{i=0}^{k_1 R_{\max}} n_{2\max}(R_1, R_2)\dfrac{(R_1^2 + R_1^4)(R_2^2 + R_2^4)}{4}$ number of iterations for loops 4-8;

$k_3 = \left(\dfrac{2R_{\max}^3 + 3R_{\max}^2 + R_{\max}}{6} + \dfrac{6R_{\max}^5 + 15R_{\max}^4 + 10R_{\max}^3 - R_{\max}}{30} - 2\right)\Big/2$ number of iterations for loops 9-11.

Now, we need to know the values of all parameters for each tread. The special preparation was made before to run the main kernel where we calculate the porosity and particle-size distribution. We created the arrays of values R_1, n_1, R_2, n_2 for each work-item for the main kernel.

All the remaining variables are determined in each work-item.

The number of solids for the first generator (loop 5) is:

$$S1 = \frac{-(2R_1^2 + 1) - \sqrt{(2R_1^2 + 1)^2 - \dfrac{4c}{k_3(R_2^2 + R_2^4)}}}{2}$$

c is the remainder from dividing the total number of possible combinations k to number of possible combinations of variable (R_1, n_1, R_2, n_2).

The number of pores for first generator (loop 6) is:

$$P_1 = c - \frac{S_1(2R_1^2 - S_1 + 1)k_3(R_2^2 + R_2^4)}{2k_3(R_2^2 + R_2^4)}$$

The number of solids for second generator (loop 7):

$$S_2 = \frac{-(2R_2^2 + 1) - \sqrt{(2R_2^2 + 1)^2 - \dfrac{8c}{k_3}}}{2}$$

The number of pores for second generator (loop 8):

$$P_2 = c - \frac{S_2(2R_1^2 - S_1 + 1)k_3}{2k_3}$$

The number of subdivisions for third generator (loop 9):

$$R_3 = \left(\frac{2(R_3^3 + 3R_3^2 + R_3)}{6} + \frac{(6R_3^5 + 15R_3^4 + 10R_3^3 - R_3)}{30} - 2 \right) \Big/ 2$$

The number of solids for third generator (loop 10):

$$S_3 = \frac{-(2R_3^2 + 1) - \sqrt{(2R_3^2 + 1)^2 - \frac{8c}{k_3}}}{2}$$

The number of pores for third generator (loop 11):

$$P_3 = c - \frac{S_3(2R_3^2 - S_3 + 1)}{2}$$

4. Results and discussion

This parallel algorithm was tested on a 6 GB, 448 cores Tesla C2070 GPU (CUDA SDK 4.0) from Nvidia [16]. For portability, we chose the OpenCL (Open Computing Language) framework [17], which offers an abstract view of the parallel architecture used. This allowed us to use both the CPU (Central Processing Units) and the GPU backend.

Comparison of computation time for CPU and GPU is presented in table 3. The CPU used is an Intel Xeon E5506 processor running at 2.13GHz. The computation time for GPU includes the runtime kernel compilation time.

Table 3. Comparison of computation time

R_{max}	Number the patterns	Time CPU	Time GPU
3	10943625	8 sec	1.5 sec
4	310388561	18 min	4.5 sec

Using this algorithm, we achieved a significant reduction of computation time. It is now possible to increase the maximal scaling factor R_{max} beyond these values and to explore even more combinations in order to select the best one within a reasonable amount of time. This was our first attempt at adapting our code to innovative parallel architectures such as GPU. We are satisfied with the results we obtained and we look for others perspectives.

A soil sample can contains billions of particles. At every time step, all of these particles interact with each other with water flowing between them. The parameters that characterize this flow are also constantly changing in response to these interactions. Thus, the number of operations needed to compute one simulated second for a single cubic centimeter of soil is immense. Our tests have shown that the method proposed here offer a good potential to adapting to parallel architectures.

Use of simulation environments such as our virtual laboratory opens new and interesting possibilities. We are currently investigating the possibility of (i) computing the motion of particles under water flow in real time, (ii) increasing of the number of particles or including finer particles, (iii) increasing the accuracy of the results for the Navier-Stokes equations, (iv) adding new physical phenomena in our model such as the thermo effects.

References

[1] Indraratna B., and Vafai F (1997) “Analytical Model for Particle Migration Within Base Soil – Filter System”, Jour. Geotech. & GeoEnv. Eng., ASCE, vol 123, No 2, pp. 100-109

[2] Witt K. (1993) “Reliability Study of Granular Filters”, Filtersin Geotech. & Hydr. Eng., Braun, Heibaum &Schuler (eds.) 1993 Balkema, Rotterdam, pp. 35-41

[3] T.C. Kenney, R.Chanal, E. Chiu, G.I. Ofoegbu, G.N. Omange and C.A. Ume (1985) “Comtroling constriction size of granular filters”, Can. Geotech. J, 22, pp. 32-43 ,

[4] Perrier, E., Bird, N., and Rieu, M. (1999) "Generalizing the fractal model of soil structure: the pore-solid fractal approach." Geoderma, Vol. 88: pp. 137-164.

[5] Mandelbrot, B.B. 1983 the fractal geometry of nature, W.H. Freeman, New York

[6] Friesen, W.I., Mikula, R.J., (1987) fractal dimensions of coal particles. J. Colloid interf. 120, pp 263-271

[7] M.A. Soto and O.M. Vilar “Evaluation of a Pore Fractal Model for the Prediction of Soil water retention Curve”, unsaturated Soils (2006), pp. 2441-2452

[8] Toledo, P.G., Novy, R.A., davis, H.T., Scriven, L.E., 1990. Hydraulic conductivity of porous media at low content. Soil Sci. Soc. Am, J. 54, 673-679

[9] Lehmann, P., Stähli, M., Papritz, A., Gygi, A., and Flühler, H. (2003). "A Fractal Approach to Model Soil Structure and to calculate thermal conductivity of soils," Transport in porous media, Vol. 52: pp. 313-332.

[10] Edited by Y. Pachepsky, J.W. Crawford, W.J. Rawls “Fractals in soil sciences” 2000, ELSEVIER, 295 p.

[11] Goldin, A. L. and Rasskazov, L. N. (1992). Earth Dam Design, *Geotechnika 2.* Balkema, Rotterdam, 328 p.

[12] L.M. Arya, F.J. Leij, P.J. Shouse and M.Th. van Genuchten “relation between the Hydraulic Conductivity function and the particle-Size distribution” (1999), Soil Sci. Soc . Am J 63: 1063-1070

[13] Chekired M. and Roubtsova V. (2010) “2-D pore-particle scale model of the erosion at the boundary of two soils under horizontal groundwater flow”, 5th International Conference on Scour and Erosion ICSE-5

[14] Scheidegger, A. E. (1963). "The Physics of Flow Through Porous Media." University of Toronto Press, Toronto.

[15] Vukovic, M. and Soro, A. (1992). "Determination of Hydraulic Conductivity of Porous Media from Grain-Size Composition." Water Resources Publications, Littleton, Colorado, CO, 83p.

[16] NVIDIA Corporation (2009) NVIDIA’s Next Generation CUDA Compute Architecture: Fermi, NVIDIA Corporation.

[17] Munshi, A. (2011) The OpenCL Specification Version 1.1, The Khronos Group. 5. Mark, S. and Akeley, K., (2010) The OpenGL Graphics System: A SpecificationVersion 4.1 (Core Profile), *The Khronos Group.*

A parallel adaptive method for pseudo-arclength continuation

D A Aruliah[1], L van Veen[1] and A Dubitski[2]

[1] Faculty of Science, University of Ontario Institute of Technology, Oshawa, ON, Canada, L1H 7K4

E-mail: [1]{Dhavide.Aruliah,Lennaert.vanVeen}@uoit.ca, [2]alex.dubitski@gmail.com

Abstract. Pseudo-arclength continuation is a well-established method for constructing a numerical curve comprising solutions of a system of nonlinear equations. In many complicated high-dimensional systems, the corrector steps within pseudo-arclength continuation are extremely costly to compute; as a result, the step-length of the preceding prediction step must be adapted carefully to avoid prohibitively many failed steps. We describe the essence of a parallel method for adapting the step-length of pseudo-arclength continuation. Our method employs several predictor-corrector sequences with differing step-lengths running concurrently on distinct processors. Our parallel framework permits intermediate results of correction sequences that have not yet converged to seed new predictor-corrector sequences with various step-lengths; the goal is to amortize the cost of corrector steps to make further progress along the underlying numerical curve. Results from numerical experiments suggest a three-fold speed-up is attainable when the continuation curve sought has great topological complexity and the corrector steps require significant processor time.

1. Introduction

Continuation (or homotopy) problems arise naturally in numerous application domains. For instance, in numerical studies of turbulent fluid flow, the Reynolds number Re can be treated as a scalar that parametrizes a family of discretized versions of the Navier-Stokes equations. Certain computations–for instance, computations of time-periodic solutions in fluid flow—are significantly less challenging at low Reynolds numbers than at high Reynolds numbers, where dynamical processes occur on a larger range of spatial scales. Continuation can be used to extend the results obtained at some low Reynolds number into the physically more interesting regime, in other words to construct a homotopy between turbulent flows with very different characteristics. Homotopies are also used to relate flows in different geometries and study their similarities and differences (see [1] and references therein).

In mathematical terms, continuation problems are nonlinear systems of equations where the number of equations is one fewer than the number of variables. That is, given a mapping $\mathbf{F} : \mathbb{R}^n \times \mathbb{R} \to \mathbb{R}^n$, the equation

$$\mathbf{F}(\mathbf{x}, \lambda) = \mathbf{0} \tag{1}$$

with vector $\mathbf{x} \in \mathbb{R}^n$ and scalar $\lambda \in \mathbb{R}$ defines a continuation problem. In geometric terms, (1) defines a one-dimensional implicit curve in $\mathbb{R}^{n+1}$ under suitable smoothness and consistency properties of the mapping $\mathbf{F}$.

Published under licence by IOP Publishing Ltd

The essential idea of continuation (sometimes called homotopy) is to follow the underlying curve of solutions. Specifically, we assume that the parameter λ lies in some interval $[\lambda_{\text{initial}}, \lambda_{\text{final}}] \subseteq \mathbb{R}$. Often, the problem can be formulated in such a way, that solving (1) for $\mathbf{x}$ is easy when $\lambda = \lambda_{\text{initial}}$, and solving (1) for $\mathbf{x}$ is challenging when $\lambda = \lambda_{\text{final}}$. Thus, continuation is the process of gradually morphing the solution of a straightforward problem into the solution of a formidable problem in small increments.

Numerical continuation refers to families of numerical algorithms for generating points on a homotopy curve. Natural continuation and pseudo-arclength continuation are examples of predictor-corrector methods [2, 3, 4]. Predictor-corrector algorithms are inherently sequential: one solution point is used to predict the next solution point on the curve. Parallelization can be introduced within the corrector steps, but the predictor steps need to be computed in sequence.

More importantly, predictor-corrector methods rely on adaptive step-size selection to make optimal progress in moving along the curve of solutions. Strategies for adapting step-sizes selection are largely heuristic: when corrector steps fail to converge, we reduce the step-size, we reject the predictor step, and we start again. This turns out to be the principal bottleneck in many continuation problems, as the computation time devoted to nonconvergent sequences of corrector steps is wasted. Thus, identifying strategies that can improve the performance of predictor-corrector schemes by reducing the cost of failed predictor steps is a significant challenge in modern High-Performance Computing.

We describe in the present work a framework for parallelizing predictor-corrector methods to amortize the cost of failed predictor steps. Specifically, we carry out several predictor steps of different step-sizes for pseudo-arclength continuation in parallel on distinct processors. Our algorithm manages the different steps in parallel using a colored rooted tree. The principal context within which this approach is practical is when users have a serial application code that takes a long time to evaluate corrector steps. If the underlying numerical curve is sufficiently complex that predictor steps fail often, our parallel adaptive strategy offers some hope of progress.

2. Background and motivation

Before describing our parallel adaptive algorithm, we briefly review previous numerical continuation algorithms—notably natural parameter and pseudo-arclength continuation—for solving problems of the form (1), where $\lambda \in [\lambda_{\text{initial}}, \lambda_{\text{final}}] \subseteq \mathbb{R}$, $\mathbf{x} \in \mathbb{R}^n$, and $\mathbf{F} : \mathbb{R}^n \times \mathbb{R} \to \mathbb{R}^n$. In doing so, we sometimes write the underlying equation (1) in the form

$$\mathbf{F}(\mathbf{z}) = \mathbf{0}, \text{ where } \mathbf{z} = (\mathbf{x}, \lambda) \in \mathbb{R}^{n+1}. \tag{2}$$

That is, we treat the concatenation of the n-vector $\mathbf{x}$ and the scalar λ as a single $(n+1)$-vector $\mathbf{z}$ while using the same symbol $\mathbf{F}$ to denote the mapping. Also, the appropriately-sized matrices representing Jacobian derivatives are denoted by

$$\mathbf{F}_\lambda \equiv \frac{\partial \mathbf{F}}{\partial \lambda} \in \mathbb{R}^{n\times 1}, \quad \mathbf{F}_\mathbf{x} \equiv \frac{\partial \mathbf{F}}{\partial \mathbf{x}} \in \mathbb{R}^{n\times n}, \quad \text{and} \quad \mathbf{F}_\mathbf{z} \equiv \frac{\partial \mathbf{F}}{\partial \mathbf{z}} \in \mathbb{R}^{n\times (n+1)} \quad \text{respectively.} \tag{3}$$

An obvious method for constructing a curve of numerical solutions of (1) is to increment the parameter λ gradually from λ_{initial} to λ_{final}. That is, given a point $(\mathbf{x}, \lambda) \in \mathbb{R}^{n+1}$ known to lie on the curve, a new point on the curve $(\boldsymbol{\xi}, \lambda + h)$ is found

(i) by incrementing λ by a small amount $h > 0$; and

(ii) by solving $\mathbf{F}(\boldsymbol{\xi}, \lambda + h) = \mathbf{0}$ for the unknown $\boldsymbol{\xi}$.

This approach is referred to as *natural parameter continuation* (or, more simply, *natural continuation*, see Algorithm 1) [2, 3, 4]. Fixing the known parameter value $\lambda + h$ in the underdetermined nonlinear system $\mathbf{F}(\boldsymbol{\xi}, \lambda + h) = \mathbf{0}$ yields a closed $n \times n$ system in the n unknown components of $\boldsymbol{\xi}$. Intuitively, if h is sufficiently small, the vector $\boldsymbol{\xi}$ should be easy to obtain using a quadratically-convergent Newton iteration (see, *e.g.*, [5]).

ALGORITHM 1: Natural parameter continuation.

Input: $[\lambda_{\text{initial}}, \lambda_{\text{final}}] \subseteq \mathbb{R}$; vector $\mathbf{x}^{(0)} \in \mathbb{R}^n$ where $\mathbf{F}(\mathbf{x}^{(0)}, \lambda_{\text{initial}}) = \mathbf{0}$; step-size $h > 0$
Output: vector $\mathbf{x}^{(k)} \in \mathbb{R}^n$ and scalar $\lambda^{(k)} \geq \lambda_{\text{final}}$ such that $\mathbf{F}\left(\mathbf{x}^{(k)}, \lambda^{(k)}\right) = \mathbf{0}$

1 $k \leftarrow 0$
2 $\lambda^{(0)} \leftarrow \lambda_{\text{initial}}$ `// initialization`
3 **while** $\lambda^{(k)} < \lambda_{\text{final}}$ **do** `// loop to generate successive points on curve`
4     $\lambda \leftarrow \lambda^{(k)} + h$ `// predictor step`
5   Iteratively solve $n \times n$ nonlinear system of equations

$$\mathbf{F}(\boldsymbol{\xi}, \lambda) = \mathbf{0}$$

    to obtain $\boldsymbol{\xi} \in \mathbb{R}^n$ starting from initial iterate $\boldsymbol{\xi}^{(0)} = \mathbf{x}^{(k)}$ `// corrector steps`
6   **if** *iteration in line 5 converges to* $\boldsymbol{\xi}$ **then** `// accept next point on curve`
7         $\mathbf{x}^{(k+1)} \leftarrow \boldsymbol{\xi}$
8         $\lambda^{(k+1)} \leftarrow \lambda$
9         $k \leftarrow k + 1$
10   **else**
11   Reduce step-size h `// reject predictor step & repeat`
12 **return** $\mathbf{x}^{(k)}$, $\lambda^{(k)}$

Natural continuation is conceptually simple and easy to implement; however, it breaks down when the curve implicitly defined by (1) admits a *fold point* (*i.e.*, a point $(\mathbf{x}, \lambda)$ where the Jacobian matrix $\mathbf{F}_{\mathbf{x}}(\mathbf{x}, \lambda)$ is singular) somewhere along the curve (see [6] for an alternative characterization of fold points). At a fold point, systematically incrementing λ as in Algorithm 1 yields an inconsistent system of nonlinear equations that cannot be solved regardless of how small h is. Fold points do occur in practical continuation problems, so other continuation strategies need to be devised [7, 8].

Pseudo-arclength continuation (as outlined in Algorithm 2) is a standard approach to circumvent fold point singularities that capitalizes on the smoothness of the underlying curve [2, 6, 9]. Under the assumption that both $\mathbf{x}$ and λ are smooth functions of arclength, pseudo-arclength continuation constitutes using the unit direction $\mathbf{T} \in \mathbb{R}^{n+1}$ tangent to the curve for prediction. The term "pseudo-arclength" applies because the step-size h is a close approximation of the Euclidean distance (in $\mathbb{R}^{n+1}$) between successive points on the numerical curve (which, in turn, is almost the arclength between the two points as measured along the curve). The tangent direction $\mathbf{T}$ can be determined by computing a null vector of the $n \times (n+1)$ Jacobian matrix $\mathbf{F}(\mathbf{x}, \lambda)$ or by computing finite differences between successive points on the curve. The underdetermined nonlinear system of equations (1) is then closed by requiring that the solution $(\mathbf{x}, \lambda)$ sought must lie in the hyperplane orthogonal to the tangent direction $\mathbf{T}$ at distance h from the last known point; see line 7 of Algorithm 2.

Natural continuation and pseudo-arclength continuation both fit into a broader framework of *predictor-corrector methods* [2]. Predictor-corrector methods involve three important components:

- a *predictor step* of a prescribed step-size;
- a sequence of iterative *corrector steps*; and
- an adaptive step-size selection strategy.

ALGORITHM 2: Pseudo-arclength continuation.

Input: $[\lambda_{\text{initial}}, \lambda_{\text{final}}] \subseteq \mathbb{R}$; vector $\mathbf{x}^{(0)} \in \mathbb{R}^n$ where $\mathbf{F}(\mathbf{x}^{(0)}, \lambda_{\text{initial}}) = \mathbf{0}$; step-size $h > 0$
Output: vector $\mathbf{x}^{(k)} \in \mathbb{R}^n$ and scalar $\lambda^{(k)} \geq \lambda_{\text{final}}$ such that $\mathbf{F}\left(\mathbf{x}^{(k)}, \lambda^{(k)}\right) = \mathbf{0}$

1 $k \leftarrow 0$
2 $\lambda^{(0)} \leftarrow \lambda_{\text{initial}}$ `// initialization`
3 $\mathbf{z}^{(0)} \leftarrow \left(\mathbf{x}^{(0)}, \lambda^{(0)}\right) \in \mathbb{R}^{n+1}$
4 Determine approximate tangent vector $\mathbf{T}^{(0)} \in \mathbb{R}^{n+1}$ at $\mathbf{z}^{(0)}$
5 **while** $\lambda^{(k)} < \lambda_{\text{final}}$ **do** `// loop to generate successive points on curve`
6     $\mathbf{w} \leftarrow \mathbf{z}^{(k)} + h\mathbf{T}^{(k)}$ `// predictor step`
7   Iteratively solve $(n+1) \times (n+1)$ nonlinear system of equations

$$\mathbf{F}(\boldsymbol{\zeta}) = \mathbf{0},$$
$$\mathbf{T}^{(k)^T}\left(\boldsymbol{\zeta} - \mathbf{z}^{(k)}\right) = h$$

    to obtain $\boldsymbol{\zeta} \in \mathbb{R}^{n+1}$ starting from initial iterate $\boldsymbol{\zeta}^{(0)} = \mathbf{w}$ `// corrector steps`
8   **if** *iteration in line 7 converges to* $\boldsymbol{\zeta}$ **then** `// accept next point on curve`
9         $\mathbf{x}^{(k+1)} \leftarrow \zeta_{1:n}$ `// extract subvector from ζ`
10         $\lambda^{(k+1)} \leftarrow \zeta_{n+1}$ `// extract last element from ζ`
11         $\mathbf{z}^{(k+1)} \leftarrow \left(\mathbf{x}^{(k+1)}, \lambda^{(k+1)}\right)$
12   Determine approximate tangent vector $\mathbf{T}^{(k+1)} \in \mathbb{R}^{n+1}$ at $\mathbf{z}^{(k+1)}$
13   Heuristically adjust the step-size h
14         $k \leftarrow k + 1$
15   **else**
16   Reduce step-size h `// reject predictor step w and repeat`
17 **return** $\mathbf{z}^{(k)} = \left(\mathbf{x}^{(k)}, \lambda^{(k)}\right)$

The predictor step is used to seed the iterative solution of a nonlinear system of equations (successive iterates being called corrector steps). There are a variety of ways in which the predictor and corrector steps can be chosen; see [2, 8]. For instance, the predictor step in natural continuation is simply incrementing the scalar λ by h; in pseudo-arclength continuation, prediction involves traversing a distance h along $\mathbf{T}$, the tangent direction to the curve. Whatever the particular predictor-corrector strategy, when the step-size h is too large, the inner corrector iterations can stagnate or diverge (perhaps because the initial predictor is too far from a solution or perhaps because the system is inconsistent). In either event, the predictor step is rejected and the step-size is reduced (according to a simple heuristic such as $h \leftarrow ch$ where c is a user-specified parameter satisfying $0 < c < 1$). Consult [2, Ch. 6] for more detailed adaptation strategies.

The structure of predictor-corrector methods leads to two significant consequences:

(i) *The most expensive part of the algorithm is computation of the corrector steps* (because each corrector step usually involves solving a linear system of equations).

(ii) *Rejected predictor steps are extremely expensive* (because numerous corrector steps are computed before deciding to reject the predictor).

Rejected predictor steps may not be prohibitively expensive for moderate sized systems amenable to dense, direct linear algebra solvers. However, as systems grow in size, the use of direct solvers to compute corrector steps becomes infeasible. For instance, Krylov subspace iterations can be applied within Newton iterations to determine the corrector steps within a continuation algorithm (a process referred to as *Newton-Krylov continuation* [10, 11]). In such large-scale problems, individual corrector steps can require hours of computation and the penalty incurred for failed predictor steps is punitive. Given that using too large a step-size can result in failed predictor steps, an obvious strategy might be to use very small step-sizes. Unfortunately, using too small a step-size impedes progress along the curve from λ_{initial} to λ_{final} (in both natural and pseudo-arclength continuation). Efficient step-size adaptivity involves trading off between these conflicting concerns while minimizing wasted computation.

Thus, we develop a parallel adaptive method for step-size selection within the predictor-corrector framework of pseudo-arclength continuation. The method we present is independent of the strategy used for the corrections steps and is designed for problems in which the correction steps are particularly expensive.

3. Outline of the parallel adaptive algorithm

There are two essential ways to achieve parallelism in the selection of step-sizes. The first is most obvious: to use a sequence of different step-sizes $t_1 h$, $t_2 h$, …, $t_W h$ for some positive scalars $\{t_k\}_{k=1}^{W}$. That is, given W processors, an initial point $\mathbf{z} \in \mathbb{R}^{n+1}$ known to satisfy $\mathbf{F}(\mathbf{z}) = \mathbf{0}$, and a unit tangent direction $\mathbf{T} \in \mathbb{R}^{n+1}$, each processor α computes a predictor step $\boldsymbol{\zeta}^{(\alpha,0)} = \mathbf{z} + t_\alpha h \mathbf{T}$ and uses that prediction to seed a sequence of corrector steps $\{\boldsymbol{\zeta}^{(\alpha,0)}, \boldsymbol{\zeta}^{(\alpha,1)}, \boldsymbol{\zeta}^{(\alpha,2)}, \dots\}$. Since the correction steps are independent of each other, no communication is required between processors except in generating the predictions. Each processor α keeps track of its own iteration counter ν_α and computes its own nonlinear residuals.

A second strategy to achieve parallelism in adapting step-sizes is to use intermediate computations to seed new predictions. That is, suppose processor α started from the predictor $\boldsymbol{\zeta}^{(\alpha,0)} = \mathbf{z}^{(\alpha)} + h_\alpha \mathbf{T}^{(\alpha)}$ and proceeds to compute a sequence of correction steps $\{\boldsymbol{\zeta}^{(\alpha,0)}, \boldsymbol{\zeta}^{(\alpha,1)}, \boldsymbol{\zeta}^{(\alpha,2)}, \dots\}$. Rather than waiting for the corrector iterations to converge, the intermediate iterates can be used to compute a normalized secant direction $\widehat{\mathbf{T}}^{(\alpha)}$ in the direction of $\boldsymbol{\zeta}^{(\alpha,k)} - \mathbf{z}^{(\alpha)}$. Assuming the corrector iterates are sufficiently close to the curve of solutions, the new secant direction $\widehat{\mathbf{T}}^{(\alpha)}$ approximates a tangent to the curve and can be handed off to another processor β to generate a new predictor point $\mathbf{z}^{(\beta)} + h_\beta \mathbf{T}^{(\beta)}$, where $\mathbf{z}^{(\beta)} = \boldsymbol{\zeta}^{(\alpha,k)}$ for some iterate k, $h_\beta = t_\ell h_\alpha$ for some $\ell \in \{1, \dots, W\}$, and $\mathbf{T}^{(\beta)} = \widehat{\mathbf{T}}^{(\alpha)}$.

To manage both these strategies, we represent each processor in use by a node in a rooted tree with the master processor at the root. We assume that the tree has a width $W > 0$ (which corresponds to the scalars t_1, t_2, …, t_W that multiply the step-length h) and a depth D (which corresponds to the number of extrapolations to carry out from intermediate calculations. Each processor has certain data associated with it and processors are allocated to the maximum width and depth of the tree.

Specifically, processor α is associated with a base point $\mathbf{z}^{(\alpha)} \in \mathbb{R}^{n+1}$, a step-size h_α, a unit tangent direction $\mathbf{T}^{(\alpha)}$, a current iteration counter ν_α, a current iterate $\boldsymbol{\zeta}^{(\alpha,\nu_\alpha)} \in \mathbb{R}^{n+1}$, a nonlinear residual $\mathbf{r}^{(\alpha,\nu_\alpha)} \in \mathbb{R}^n$, and a color c_α where $c_\alpha \in \{\text{GREEN}, \text{YELLOW}, \text{RED}, \text{BLACK}\}$. In the preceding list, the nonlinear residual satisfies

$$\mathbf{r}^{(\alpha,\nu_\alpha)} = \mathbf{F}\left(\boldsymbol{\zeta}_{\boldsymbol{\alpha}}^{(\nu_\alpha)}; \mathbf{z}_\alpha, \mathbf{T}_\alpha, h_\alpha\right), \quad \text{where} \quad \mathbf{G}(\boldsymbol{\zeta}; \mathbf{z}, \mathbf{T}, h) := \begin{bmatrix} \mathbf{F}(\boldsymbol{\zeta}) \\ \mathbf{T}^T(\boldsymbol{\zeta} - \mathbf{z}) - h \end{bmatrix}. \tag{4}$$

The color of each node α is assigned using the 2-norm of the residual at the current corrector step, *i.e.*, $\left\|\mathbf{r}_\alpha^{(\alpha,\nu_\alpha)}\right\|_2$. Given a positive tolerance TOL, the node α is assigned a color c_α as follows:

- c_α = GREEN if the current iterate $\boldsymbol{\zeta}^{(\alpha,\nu_\alpha)}$ is deemed to be sufficiently near to the curve of solutions, *i.e.*, $\left\|\mathbf{r}_\alpha^{(\alpha,\nu_\alpha)}\right\|_2 \leq \text{TOL}$;
- c_α = YELLOW if the next iterate $\boldsymbol{\zeta}^{(\alpha,\nu_\alpha+1)}$ is expected to be sufficiently near to the curve of solutions, *i.e.*, $\log\left\|\mathbf{r}_\alpha^{(\alpha,\nu_\alpha)}\right\|_2 \leq \frac{1}{2}\log(\text{TOL})$;

- c_α = BLACK if the iteration sequence is deemed to have diverged; and
- c_α = RED otherwise.

The specific criterion for detecting divergence of a sequence of corrector iterates depends on another user-specified parameter due to problem-specific challenges in inferring divergence. The criterion for coloring a node yellow is based on the assumption that the corrector step converges quadratically. In case the corrector steps are convergent of a different order—for instance, linearly convergent fixed-point iterations—this criterion can readily be changed. This might also be necessary for strongly nonlinear problems or for systems with a lot of numerical noise; in either of these cases, the quadratic convergence of Newton iteration is realized only very close to the solution. New processes are assigned the color red by default until the nonlinear residual is computed and the color can be reassigned.

With the preceding ideas in mind, it is possible to describe the underlying parallel adaptive algorithm. Algorithm 3 is a high level summary of the essential steps. Notice that the details of tree traversal and the precise heuristic for pruning redundant subtrees are not included in the high level algorithm description; the former is a standard recursive procedure and the latter involves comparing pseudo-arclengths of paths from the root node to the leaves. The precise details will be described in a forthcoming paper to accompany a public release of the software.

ALGORITHM 3: Essential parallel adaptive algorithm

Input: $[\lambda_{\text{initial}}, \lambda_{\text{final}}] \subseteq \mathbb{R}$; vector $\mathbf{x}^{(0)} \in \mathbb{R}^n$ where $\mathbf{F}(\mathbf{x}^{(0)}, \lambda_{\text{initial}}) = \mathbf{0}$; step-size $h > 0$; tangent direction $\mathbf{T}^{(0)} \in \mathbb{R}^{n+1}$ where $\|\mathbf{T}^{(0)}\|_2 = 1$; tolerance TOL; depth $D > 0$; width $W > 0$; positive scaling parameters $\{t_1, t_2, \ldots, t_W\}$

Output: vector $\mathbf{x}^{(k)} \in \mathbb{R}^n$ and scalar $\lambda^{(k)} \geq \lambda_{\text{final}}$ such that $\mathbf{F}\left(\mathbf{x}^{(k)}, \lambda^{(k)}\right) = \mathbf{0}$

Seed root node α with data $(\mathbf{z}^{(\alpha)}, \mathbf{T}^{(\alpha)}, h_\alpha) \leftarrow ((\mathbf{x}^{(0)}, \lambda_{\text{initial}}), \mathbf{T}^{(0)}, h)$ `// initialization`

repeat `// loop to generate successive points on curve`

  foreach *leaf node* α **do**

    if *depth of* $\alpha < D$ **then**

      $\widehat{\mathbf{T}}^{(\alpha)} \leftarrow \left\|\boldsymbol{\zeta}^{(\alpha,\nu_\alpha)} - \mathbf{z}^{(\alpha)}\right\|_2^{-1} \left(\boldsymbol{\zeta}^{(\alpha,\nu_\alpha)} - \mathbf{z}^{(\alpha)}\right)$ `// secant direction`

      for $\ell = 1 : W$ **do**

        if *processors are available* **then**

          Assign processor β_ℓ the data $\left(\mathbf{z}^{(\beta_\ell)}, \mathbf{T}^{(\beta_\ell)}, h_{\beta_\ell}, \nu_{\beta_\ell}, \boldsymbol{\zeta}^{(\beta_\ell,\nu_{\beta_\ell})}, c_{\beta_\ell}\right)$ according to

$$\mathbf{z}^{(\beta_\ell)} \leftarrow \mathbf{z}^{(\alpha)}, \mathbf{T}^{(\beta_\ell)} \leftarrow \widehat{\mathbf{T}}^{(\alpha)}, h_{\beta_\ell} \leftarrow t_\ell h_\alpha, \nu_{\beta_\ell} \leftarrow 0,$$
$$\boldsymbol{\zeta}^{(\beta_\ell,\nu_{\beta_\ell})} \leftarrow \mathbf{z}^{(\alpha)} + t_\ell h_\alpha \widehat{\mathbf{T}}^{(\alpha)}, c_{\beta_\ell} \leftarrow \text{RED}$$

  Compute single corrector steps concurrently on all RED and YELLOW nodes of the tree.

  Traverse nodes of tree updating residuals

$$\mathbf{r}^{(\alpha,\nu_\alpha)} \leftarrow \mathbf{F}\left(\boldsymbol{\zeta}_\alpha^{(\nu_\alpha)}; \mathbf{z}_\alpha, \mathbf{T}_\alpha, h_\alpha\right)$$

  and updating colors c_α accordingly.

  Prune the tree, deleting BLACK nodes and eliminating redundant subtrees.

until *tree consists of solitary node* α *with* $\lambda^{(\alpha)} \leftarrow \mathbf{e}_{n+1}^T \boldsymbol{\zeta}^{(\alpha,\nu_\alpha)} \geq \lambda_{\text{final}}$

4. Test case: a modified Kuramoto-Sivashinsky equation

To illustrate the application of Algorithm 3 in practice, we compute a family of solutions of a modified Kuramoto-Sivashinsky equation. Specifically, we consider the PDE

$$u_t + uu_x + \nu u_{xxxx} - A\sin(u) = 0, \qquad x \in (0, 2\pi),\ t \geq 0, \tag{5a}$$

$$u^{(k)}(0,t) \equiv u^{(k)}(2\pi, t), \qquad k \in \{0, 1, 2, 3, 4\}, \tag{5b}$$

where the solutions $u = u(x,t)$ sought are expected to be at least 4 times continually differentiable on the periodic one-dimensional domain $(0, 2\pi)$. This equation has a reflection symmetry given by $S : (u, x) \mapsto (-u, -x)$ and a continuous translational symmetry. We fix $A = 8$ and consider $\nu > 0$ as homotopy parameter.

For all values of the control parameter ν, there are trivial solutions at $u = q\pi$ for $q \in \mathbb{Z}$. The equilibria with even q are unstable and those with odd q are stable for $\nu < 1/(4A)$. Beyond that point, stable, spatially varying equilibrium solutions arise. From these equilibria, several branches of travelling wave solutions bifurcate. One such branch is shown in Fig. 1. This branch has a number of fold points and the travelling wave changes its shape rapidly along the branch. Therefore, the traditional pseudo-arclength continuation approach will suffer from many failed steps. Moreover, the wave becomes increasingly localized for small values of the control parameter. In order to resolve it correctly, we need to use fine spatial discretization which will result in time-consuming corrector steps.

We compute the travelling waves as $u(x,t) = w(x - ct)$, which results in the following Boundary Value Problem (BVP):

$$-cw' + ww' + w'' + \nu w'''' - A\sin(w) = 0; \quad w(0) = w(2\pi).$$

The linear terms in this equation are efficiently computed in Fourier space, while the nonlinear terms are best computed on a regular, periodic grid. Thus, we approximate solutions as

$$w_j = \sum_{k=0}^{N-1} a_k e^{ikx_j}, \text{ where } x_j = \frac{2\pi j}{N},\ j = 0, \ldots, N-1$$

and switch between the representations $\{w_j\}$ and $\{a_k\}$ by the discrete Fourier transform, implemented using using the FFTW library. The computational cost of each transform is $O(N \ln N)$. The resolution is set by the parameter N, fixed to $N = 2048$ to resolve the solutions shown in Fig. 1 and avoid aliasing issues.

The continuation problem now takes the form $\mathbf{F}(a_0, \ldots, a_{N-1}, c, \nu) = 0$, where $\mathbf{F}$ and its derivatives are evaluated using the pseudo-spectral method outlined above. Since we have an extra unknown, the wave speed c, we must add an extra equation to ensure uniqueness of the corrector steps. We impose that the Newton update steps be orthogonal to w_x, the generator of the symmetry group of translations. This condition ensures that the successive iterates under corrector steps so not slide along the x-direction. Because of the second nonlinear term in Eq. 5, the Jacobian matrix $\mathbf{F_x}$ is dense. We employ a direct solver to find the Newton corrector steps, so that the computational complexity of each corrector step is $O(N^3)$.

The wall-clock time for the computation of the entire curve shown in Fig. 1 using various tree structures and step size distributions is shown in Fig. 2. The maximal speed-up is about a factor of three, obtained with a tree depth and width of three. In that computation, there is a lot of redundancy as many processors will be working on very similar approximate solutions. A speed-up of more than a factor of two, however, can be obtained using as few as three processors with the same step size multiplication factor.

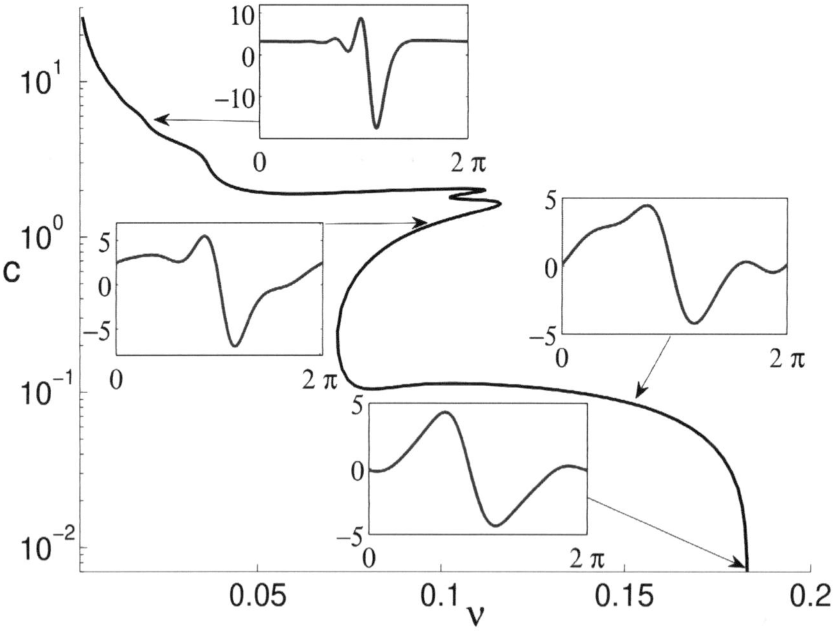

Figure 1. The branch of travelling wave solutions to Eq.5 used for testing the parallel continuation algorithm. Shown is the wave speed, c, versus the control parameter, ν. The inlays show snap shots of the solution $u(x, t)$ at four points along the continuation curve. At $c = 0$, the wave bifurcates from an equilibrium, which is symmetric under the reflection symmetry. For $\nu \lesssim 0.01$ the solution becomes strongly localized.

5. Summary

The algorithm described here is designed with nonlinear continuation problems in mind where the relative cost of failed corrector steps is high and the cost of processors is low. The situation we have in mind is where a user might have a serial code and several cores on their desktop workstation with no obvious means of parallelization. The method developed here does not exhibit good scaling as desired in High-Performance Computing applications employing hundreds or thousands of processors. However, given that most machines now have four to eight cores, a user with minimal HPC experience might employ the strategies described here to make some improvements on the serial performance of their codes. A more detailed paper and a public release of the software is currently under development.

References

[1] Kawahara G, Uhlmann M and van Veen L 2012 *Ann. Rev. Fluid Mech.* **44** 203–225
[2] Allgower E L and Georg K 2003 *Introduction to Numerical Continuation Methods* (*Classics in Applied Mathematics* vol 45) (Philadelphia: SIAM)
[3] Govaerts W 2000 *Numerical Methods for Bifurcations of Dynamic Equilibria* (Philadelphia: SIAM)
[4] Kuznetsov Y 1998 *Elements of Applied Bifurcation Theory* (New York: Springer-Verlag)
[5] Kelley C 2003 *Solving Nonlinear Equations with Newton's Method* Fundamentals of Algorithms (Philadelphia: SIAM)
[6] Dickson K, Kelley C, Ipsen I and Kevrekidis I 2007 *SIAM J. Numer. Anal.* **45** 263–276

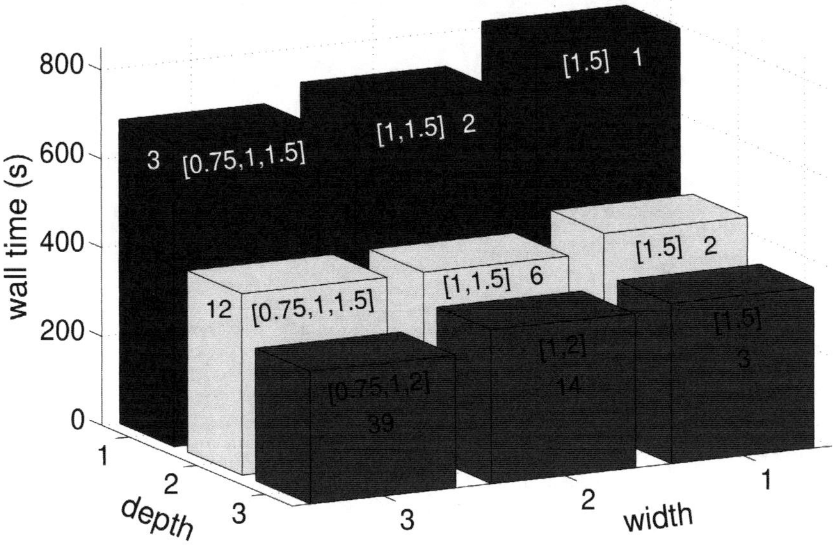

Figure 2. Wall-clock time for the computation of the continuation curve shown in Fig. 1 using a tree width and depth up to three. The numbers between brackets denote the multipliers for the step size, $\{t_i\}_{i=1}^{W}$ and the integer denotes the number of CPUs concurrently processing corrector steps.

[7] Yang Z H and Keller H 1986 *SIAM J. Sci. Stat. Comput.* **7** 351–361
[8] Doedel E J 2007 *Numerical Continuation Methods for Dynamical Systems* ed Krauskopf B, Osinga H M and Galàn-Vioque J (Springer Netherlands)
[9] Keller H 1977 *Applications of Bifurcation Theory* ed Rabinowitz P (New York: Academic Press) pp 359–384
[10] Sánchez J, Net M, García Archilla B and Simó C 2004 *J. Comp. Phys.* **201** 13–33
[11] Knoll D and Keyes D 2004 *J. Comp. Phys.* **193** 357–397

Journal of Physics: Conference Series **385** (2012) 012009 doi:10.1088/1742-6596/385/1/012009

Convergence order vs. parallelism in the numerical simulation of the bidomain equations

Oluwaseun Sharomi and Raymond J. Spiteri

Department of Computer Science, University of Saskatchewan, Saskatoon, SK, S7N 5C9, Canada

E-mail: spiteri@cs.usask.ca

Abstract. The propagation of electrical activity in the human heart can be modelled mathematically by the bidomain equations. The bidomain equations represent a multi-scale reaction-diffusion model that consists of a set of ordinary differential equations governing the dynamics at the cellular level coupled with a set of partial differential equations governing the dynamics at the tissue level. Significant computation is generally required to generate clinically useful data from the bidomain equations. Contemporary developments in computer architecture, in particular multi- and many-core computers and graphics processing units, have made such computations feasible. However, the zeal to take advantage to parallel architectures has typically caused another important aspect of numerical methods for the solution of differential equations to be overlooked, namely the convergence order. It is well known that higher-order methods are generally more efficient than lower-order ones when solutions are smooth and relatively high accuracy is desired. In these situations, serial implementations of high-order methods may remain surprisingly competitive with parallel implementations of low-order methods. In this paper, we examine the effect of order on the numerical solution of the bidomain equations in parallel. We find that high-order methods, in particular high-order time-integration methods with relatively better stability properties, tend to outperform their low-order counterparts, even when the latter are run in parallel. In other words, increasing integration order often trumps increasing available computational resources, especially when relatively high accuracy is desired.

1. Introduction

Computer simulation of human organs has the potential to revolutionize medicine as we know it, from improving our understanding of pathologies to developing innovative treatments *in silico*. The bidomain equations are a popular mathematical model for the propagation of electrical activity in myocardial tissue [1].

The complexity of the mathematical model and geometry of the tissue makes solving the bidomain equations a challenging computational problem. The heart has approximately 10 billion muscle cells. To obtain clinically useful data from the bidomain equations, they must be solved with many millions of unknowns. To obtain such data in real time is the ultimate goal; at present we are still a factor of 40 or so away from being able to do so [2, 3].

A number of software packages can be used to solve bidomain model, e.g., the Cancer, Heart, and Soft-Tissue Environment (CHASTE), Control of Cardiac Arrhythmias (CARP), OPENCMISS, etc.; see [4] and references therein. They are mainly based on low-order numerical methods. In fact, many of the software packages compared in [4] implement low-order methods to solve the bidomain model.

Published under licence by IOP Publishing Ltd

It is well known that higher-order methods are generally more efficient than low-order ones when computing (smooth) solutions to relatively high precision. However it seems that in the attempts to take advantage of parallelism, many software packages overlook the importance of higher order in the numerical methods implemented. We provide a quantitative evaluation of the effectiveness of high-order methods run in serial versus low-order methods run in parallel.

The remainder of this paper is organized as follows. In Section 2, we detail the mathematical formulation of the bidomain equations. In Section 3, we outline the software used to perform our investigation, including the underlying numerical methods. In Section 4, we describe the numerical experiments performed and present the results. In Section 5, we give our conclusions.

2. The Bidomain Equations

The electrophysiological behaviour of the heart can be mathematically modelled using a multi-scale reaction-diffusion model based on the bidomain equations [1]. In particular, the electrical activity and ionic currents of a single heart cell can be described by a system of ordinary differential equations (ODEs). These ODEs are coupled with a system of partial differential equations (PDEs) in order to model the propagation of the electrical activity throughout the entire heart via the bidomain equations. The bidomain equations are formulated as

$$\chi\Big(C_m\frac{\partial V_m}{\partial t}+I_{ion}(\mathbf{s},V_m)\Big)=\Big(\nabla\cdot(\mathbf{M}_I\nabla V_m)+\nabla\cdot(\mathbf{M}_I\nabla u_E)\Big), \tag{1a}$$

$$\nabla\cdot((\mathbf{M}_I+\mathbf{M}_E)\nabla u_E)+\nabla\cdot(\mathbf{M}_I\nabla V_m)=0, \tag{1b}$$

$$\frac{\partial \mathbf{s}}{\partial t}=\mathbf{f}(\mathbf{s},V_m), \tag{1c}$$

with boundary conditions

$$\begin{aligned}\hat{\mathbf{n}}\cdot(\mathbf{M}_I\nabla(V_m+u_E))&=0,\\ \hat{\mathbf{n}}\cdot(\mathbf{M}_E\nabla u_E)&=0,\end{aligned} \tag{1d}$$

where V_m is the transmembrane potential, u_E is the extracellular potential, $\mathbf{s}$ is a vector of state variables arising from the cell model, $\mathbf{M}_I$ and $\mathbf{M}_E$ are intracellular and extracellular conductivity tensors, respectively, χ is the area of cell membrane per unit volume, C_m is the capacitance per unit area across the cell membrane, and $I_{ion}(\mathbf{s},V_m)$ is the ionic current per unit area provided by the cell model. The function $\mathbf{f}(\mathbf{s},V_m)$ is a vector-valued function that is determined by the particular cell model used.

3. Software Description

The purpose of this study is to examine the interaction of convergence order and parallelism when solving the bidomain equations. Accordingly we use two software packages that are capable of solving the bidomain equations: one that uses low-order discretizations but can take advantage of parallelization and one that can take advantage of high-order discretizations but runs in serial.

CHASTE is a general-purpose simulation package aimed at multi-scale, computationally demanding problems arising in biology and physiology [5]. CHASTE is a parallel solver that implements low-order finite element methods for the spatial discretization and a first-order semi-implicit method, in which the linear diffusion terms in equation (1a) and (1b) are treated implicitly and the reaction terms explicitly [6], for the time integration of the PDEs. This leads to a discretization of the form

$$\begin{bmatrix}\frac{\chi C_m}{\Delta t}\mathbf{M}+\mathbf{A}_I & \mathbf{A}_I\\ \mathbf{A}_I & \mathbf{A}_I+\mathbf{A}_E\end{bmatrix}\begin{bmatrix}\mathbf{V}^n\\ \mathbf{u}_E^n\end{bmatrix}=\begin{bmatrix}\frac{\chi C_m}{\Delta t}\mathbf{M}\mathbf{V}^{n-1}\\ 0\end{bmatrix},$$

where Δt is the time step, $\mathbf{V}^n$, $\mathbf{u}_E^n$ are vectors containing the values of V_m, u_E in the domain at time $n\Delta t$, respectively, and $\mathbf{A}_I$ and $\mathbf{A}_E$ are the spatial discretizations of the operators $\nabla \cdot (\mathbf{M}_I \nabla)$ and $\nabla \cdot (\mathbf{M}_E \nabla)$, respectively [6]. The ODEs in equation (1c) are then solved using any time-integration scheme, typically an explicit method such as forward Euler or Heun's method.

On the other hand, Nektar++ is a software package designed to provide a toolbox of data structures and algorithms that implement the spectral/*hp* element method [7]. The spectral/*hp* element method can be considered to be a high-order extension of the traditional "finite" or "spectral" element methods, where convergence is not only achievable through reducing the mesh size h but also through increasing the local polynomial order P of the basis functions used to expand the solution. It combines the geometric flexibility of finite elements with the high accuracy of spectral methods [8]. At present, Nektar++ is a serial solver that uses high-order time-integration methods in the form of general linear methods (GLMs) [9].

GLMs are the unified generalization of the two principal types of time-integration schemes: linear multi-step methods and linear multi-stage (Runge–Kutta) methods. Nektar++ uses implicit-explicit (IMEX) GLMs [9], a modification of the standard GLM formulated by Burrage and Butcher [10]. Nektar++ applies the method of lines to the PDEs to yield a large coupled system of ODEs [9]. IMEX GLMs are most useful when these ODEs can be written in the form

$$\frac{d\mathbf{y}}{dt} = \mathbf{f}(\mathbf{y}) + \mathbf{g}(\mathbf{y}), \quad \mathbf{y}(t_0) = \mathbf{y}_0,$$

where $\mathbf{f} : \mathbb{R}^N \to \mathbb{R}^N$ and $\mathbf{g} : \mathbb{R}^N \to \mathbb{R}^N$ are functions with different stiffness properties [9]. Time step n of an IMEX-GLM that uses r past steps and s stages is formulated as

$$\begin{aligned} \mathbf{Y}_i &= \Delta t \sum_{j=1}^{s} a_{ij}^{IM} \mathbf{G}_j + \Delta t \sum_{j=1}^{s} a_{ij}^{EX} \mathbf{F}_j + \sum_{j=1}^{r} u_{ij} \mathbf{y}_j^{[n-1]}, \quad \text{for } i = 1 \ldots s, \\ \mathbf{y}_k^{[n]} &= \Delta t \sum_{j=1}^{s} b_{kj}^{IM} \mathbf{G}_j + \Delta t \sum_{j=1}^{s} b_{kj}^{EX} \mathbf{F}_j + \sum_{j=1}^{r} v_{kj} \mathbf{y}_j^{[n-1]}, \quad \text{for } k = 1 \ldots r, \end{aligned} \tag{2}$$

where the $\mathbf{Y}_i$ are called the stage values, the $\mathbf{F}_i$ and $\mathbf{G}_i$ are called the stage derivatives, defined as

$$\mathbf{F}_i = \mathbf{f}(\mathbf{Y}_i), \qquad \mathbf{G}_i = \mathbf{g}(\mathbf{Y}_i),$$

and the superscripts IM and EX denote *implicit* and *explicit*, respectively. The matrices $\mathbf{A}^{IM} = [a_{ij}^{IM}]$, $\mathbf{A}^{EX} = [a_{ij}^{EX}]$, $\mathbf{U} = [u_{ij}]$, $\mathbf{B}^{IM} = [b_{kj}^{IM}]$, $\mathbf{B}^{EX} = [b_{kj}^{EX}]$ and $\mathbf{V} = [v_{kj}]$ define the specific time-integration method via the $(s + r) \times (2s + r)$ block-partitioned matrix

$$\begin{bmatrix} \mathbf{A}^{IM} & \mathbf{A}^{EX} & \mathbf{U} \\ \mathbf{B}^{IM} & \mathbf{B}^{EX} & \mathbf{V} \end{bmatrix}.$$

Letting $\mathbf{Y} = \begin{bmatrix} \mathbf{Y}_1 \\ \mathbf{Y}_2 \\ \vdots \\ \mathbf{Y}_s \end{bmatrix}$, $\mathbf{F} = \begin{bmatrix} \mathbf{F}_1 \\ \mathbf{F}_2 \\ \vdots \\ \mathbf{F}_s \end{bmatrix}$, $\mathbf{G} = \begin{bmatrix} \mathbf{G}_1 \\ \mathbf{G}_2 \\ \vdots \\ \mathbf{G}_s \end{bmatrix}$, $\mathbf{y}^{[n-1]} = \begin{bmatrix} \mathbf{y}_1^{[n-1]} \\ \mathbf{y}_2^{[n-1]} \\ \vdots \\ \mathbf{y}_r^{[n-1]} \end{bmatrix}$, and $\mathbf{y}^{[n]} = \begin{bmatrix} \mathbf{y}_1^{[n]} \\ \mathbf{y}_2^{[n]} \\ \vdots \\ \mathbf{y}_r^{[n]} \end{bmatrix}$,
equation (2) can be re-written in tensor form as

$$\begin{bmatrix} \mathbf{Y} \\ \mathbf{y}^{[n]} \end{bmatrix} = \begin{bmatrix} \mathbf{A}^{IM} \otimes I_N & \mathbf{A}^{EX} \otimes I_N & \mathbf{U} \otimes I_N \\ \mathbf{B}^{IM} \otimes I_N & \mathbf{B}^{EX} \otimes I_N & \mathbf{V} \otimes I_N \end{bmatrix} \begin{bmatrix} \Delta t\, \mathbf{G} \\ \Delta t\, \mathbf{F} \\ \mathbf{y}^{[n-1]} \end{bmatrix},$$

where I_N is an $N \times N$ identity matrix and $\bigotimes$ denotes the Kronecker product. A number of IMEX GLMs are implemented in Nektar++. In our experiments reported below, we use methods from [11], denoted by ARS$(\sigma, \hat{\sigma}, p)$ for an IMEX-RK method with σ implicit stages, $\hat{\sigma}$ explicit stages, and order p.

4. Numerical Experiments

We begin by introducing the criteria used to quantify the effectiveness of an individual solver. The effectiveness of a solver is measured in terms of the accuracy achieved. For each example, we have computed a highly accurate reference solution, i.e., a solution to the bidomain problem that is known to have converged to d digits of accuracy at M points in space and time, where d is sufficiently large and determined by comparing increasingly accurate solutions and counting the number of matching digits for all M points. We use 21 equally spaced points in time for all examples considered. We use $N = 101$ points in space in the one-dimensional problems so that $M = 101 \times 21 = 2,121$. We use $N = 101 \times 101 = 10,201$ points in space for the two-dimensional problem so that $M = 10,201 \times 21 = 214,221$.

We have used both software packages to solve three different examples with different levels of accuracy. The examples are a one-dimensional problem with a periodically varying solution, a one-dimensional problem taken from [12] of a propagating wavefront, and a two-dimensional problem with a periodically varying solution; details are given below. All reference solution were computed in Nektar++. For the example of the periodically varying solution in one dimension, we have computed a reference solution with 4 matching digits using $N = 701$ mesh points, polynomial degree $P = 20$, and the semi-implicit backward differentiation formula (SBDF3) from [13] with $\Delta t = 2\text{e}{-6}$. For the example of propagating wavefront in one dimension, we have computed a reference solution with 4 matching digits using $N = 701$ mesh points, polynomial degree $P = 10$, and SBDF3 with $\Delta t = 1\text{e}{-6}$. For the example of the periodically varying solution in two dimensions, we have computed a reference solution with 10 matching digits using $N = 101 \times 101 = 10,201$ mesh points, polynomial degree $P = 9$, and ARS(1,2,1) with $\Delta t = 5\text{e}{-5}$. The accuracy of a given method is determined by computing the largest Δt and the smallest N and P that yield the required accuracy; the results are presented in Table 1.

In CHASTE, the problems were solved using shared- and distributed-memory architectures. In the shared-memory architecture, we use a single computer node with multiple processes. In the distributed-memory architecture, we use multiple computer nodes with a single process per node. A hybrid-memory architecture consisting of 8 nodes and 8 processes per node was found to underperform and hence results are omitted.

Example 1: One-dimensional problem with periodically varying solution
Consider (1) over the one-dimensional spatial domain $[0, 70]$ with $C_m = 1\ \mu\text{F/cm}^2$, $\chi = 1400\ \text{cm}^{-1}$, $\mathbf{M}_I = \mathbf{M}_E = 1$ mS/cm, $I_{ion}(\mathbf{s}, V_m) = -g_1(s, V_m)$, $\mathbf{f}(\mathbf{s}, V_m) = f_1(s, V_m)$, from $T = 0$ ms to $T = 30$ ms, the FitzHugh–Nagumo cell model [14]

$$f_1(s, V_m) = 10\left(V_m - \frac{V_m^3}{3} - s\right), \quad g_1(s, V_m) = 0.1(V_m + 1 - 0.5s), \tag{3}$$

and initial condition

$$V_m(x, y, 0) = -1.28791 + \sin(x), \ \ u_E(x, y, 0) = 0, \ \text{ and } \ s(x, y, 0) = -0.5758.$$

We have used the normalization $u_E(0, t) = 0$. The problem is solved using CHASTE and Nektar++ using the parameters in Table 1 and the shortest execution times are given in Table 2. The results show that Nektar++ outperforms CHASTE in achieving one matching digit of accuracy. This problem is sufficiently small that effectively no parallel scaling is achieved.

Table 1. Time step Δt, in milliseconds, PDE solver, ODE solver, number of mesh points N, and polynomial degree P used in CHASTE and Nektar++ for Examples 1, 2, and 3.

Example 1					
One matching digit					
Software	Δt	PDE Solver	ODE Solver	N	P
CHASTE	4.000e−3	Semi-implicit	Heun	301	—
Nektar++	6.000e−2	ARS(2,3,3)	ARS(2,3,3)	101	4
Three matching digits					
CHASTE	1.875e−4	Semi-implicit	Heun	1, 701	—
Nektar++	7.500e−4	ARS(2,3,3)	ARS(2,3,3)	11	40
Example 2					
One matching digit					
CHASTE	1.000e−3	Semi-implicit	Heun	1, 201	—
Nektar++	6.250e−3	ARS(1,2,1)	ARS(1,2,1)	101	6
Three matching digits					
CHASTE	1.000e−5	Semi-implicit	Heun	26, 101	—
Nektar++	1.000e−4	ARS(1,2,1)	ARS(1,2,1)	101	21
Example 3					
Two matching digits					
CHASTE	5.000e−4	Semi-implicit	Heun	40, 401	—
Nektar++	1.875e−4	ARS(1,2,1)	ARS(1,2,1)	10, 201	3
Ten matching digits					
CHASTE	—	Semi-implicit	Heun	1, 002, 001	—
Nektar++	5.000e−5	ARS(1,2,1)	ARS(1,2,1)	10, 201	6

The results further show that when three matching digits are required, Nektar++ outperforms CHASTE on all architectures except on a shared-memory architecture with at least 8 processes.

Table 2. Shortest execution times, in seconds, for CHASTE and Nektar++ to give one and three matching digits of accuracy for the one-dimensional problem with periodically varying solution.

One matching digit					
Shared memory					
Software	1 Procs	2 Procs	4 Procs	8 Procs	12 Procs
CHASTE	1.81316	1.62713	1.63186	1.98794	3.01474
Distributed memory					
	1 Procs	2 Procs	8 Procs	12 Procs	32 Procs
	1.81316	10.72050	30.97580	42.0600	83.19090
Nektar++	**0.90255**				
Three matching digits					
Shared memory					
	1 Procs	2 Procs	4 Procs	8 Procs	12 Procs
CHASTE	164.57200	94.95260	58.54150	42.59510	**41.02530**
Distributed memory					
	1 Procs	2 Procs	8 Procs	12 Procs	32 Procs
	164.57200	199.34400	373.85700	454.58800	686.70000
Nektar++	54.97220				

Example 2: One-dimensional problem with propagating wavefront solution
Consider (1) over the one-dimensional spatial domain $[0, 70]$ with $\chi = 1\ \text{cm}^{-1}$, $I_{ion}(\mathbf{s}, V_m) = -g_1(s, V_m)$, $\mathbf{f}(\mathbf{s}, V_m) = f_1(s, V_m)$, other parameters as in Example 1, from $T = 0$ ms to $T = 30$

ms, and initial condition given by

$$
\begin{aligned}
V_m(x,y,0) &= \begin{cases} 2, & \text{if } x \leq 3.5, \\ -1.28791, & \text{if } x > 3.5, \end{cases} \\
u_E(x,y,0) &= 0, \\
s(x,y,0) &= -0.5758.
\end{aligned}
$$

The problem is solved using CHASTE and Nektar++ using the parameters in Table 1 and the shortest execution times are given in Table 3. The results show that Nektar++ outperforms CHASTE in all situations considered here. We observe similar results for a two-dimensional version of this example described in [12].

Table 3. Shortest execution times, in seconds, for CHASTE and Nektar++ to give one and three matching digits of accuracy for the one-dimensional problem with propagating wavefront solution.

One matching digit					
Shared memory					
Software	1 Procs	2 Procs	4 Procs	8 Procs	12 Procs
CHASTE	22.32070	13.32200	8.56333	6.47482	6.71855
Distributed memory					
	1 Procs	2 Procs	8 Procs	12 Procs	32 Procs
	22.32070	32.53680	61.24480	70.74510	105.31400
Nektar++	**5.50030**				
Three matching digits					
Shared memory					
	1 Procs	2 Procs	4 Procs	8 Procs	12 Procs
CHASTE	47863.10000	24303.50000	12873.10000	7385.85000	5613.08000
Distributed memory					
	1 Procs	2 Procs	8 Procs	12 Procs	32 Procs
	47863.10000	25317.50000	12178.60000	12059.60000	13867.30000
Nektar++	**763.26600**				

Example 3: Two-dimensional problem with periodically varying solution
Consider (1) over the two-dimensional spatial domain $[0,1] \times [0,1]$ with $\mathbf{M}_I = 1.75\,\mathbf{I}_2$ mS/cm, $\mathbf{M}_E = 7\,\mathbf{I}_2$ mS/cm, $I_{ion}(\mathbf{s}, V_m) = -g_2(s, V_m)$, $\mathbf{f}(\mathbf{s}, V_m) = f_2(s, V_m)$, other parameters as in Example 1, from $T = 0$ ms to $T = 5$ ms, using the modified Fitzhugh–Nagumo cell model [14, 15]

$$
\begin{aligned}
g_2(s, V_m) &= -0.0004(V_m + 85)[s + (V_m + 70)(V_m - 40)], \\
f_2(s, V_m) &= 0.63(V_m + 85) - 0.013s,
\end{aligned}
\tag{4}
$$

and initial condition given by

$$
V_m(x,y,0) = -85 + 100(1 - \sin(xy)), \quad u_E(x,y,0) = 0, \quad \text{and} \quad s(x,y,0) = 0.
$$

We have used the normalization $u_E(0,t) = 0$. The problem is solved using CHASTE and Nektar++ using parameters in Table 1 and the shortest execution times are given in Table 4. The results show that CHASTE outperforms Nektar++ for two matching digits of accuracy when two or more processes are used in either shared or distributed memory architectures. However, Nektar++ outperforms again when more accuracy is required. Specifically Nektar++ achieved ten matching digits of accuracy just under 2.5 days of execution time whereas CHASTE requires more than 4 weeks in either shared or distributed architectures.

Table 4. Shortest execution times, in seconds, for CHASTE and Nektar++ to give two and ten matching digits of accuracy for the two-dimensional problem with periodically varying solution.

Two matching digits					
Shared memory					
	1 Procs	2 Procs	4 Procs	8 Procs	12 Procs
CHASTE	6409.01000	3408.92000	1841.12000	1063.14000	**831.07800**
Distributed memory					
	1 Procs	2 Procs	8 Procs	12 Procs	32 Procs
	6409.01000	3478.35000	1353.00000	1300.58000	1381.51000
Nektar++	4166.13000				
Ten matching digits					
Shared or Distributed memory					
	1 Procs	2 Procs	4 Procs	8 Procs	12 Procs
CHASTE	> 4 weeks	> 4 weeks	> 4 weeks	> 4 weeks	> 4 weeks
Nektar++	**205596.34200**				

5. Conclusions

CHASTE and Nektar++ have been used to solve the bidomain equations (1) in three different scenarios. The results show that high-order methods run in serial can outperform low-order methods run in parallel, especially when relatively high accuracies are desired. There are two main reasons for this behaviour. First, higher-order methods are more efficient than lower-order methods when smooth solutions are to be computed to relatively high accuracies because of their increased rate of error reduction relative to their increased expense. Second, current computational methods for the numerical solution of time-dependent differential equations typically do not scale well past a relatively small number of processes. Accordingly, there is a fairly tight limit on the time savings that can be achieved by increasing the number of processes used. These findings suggest software developers for the bidomain equations (and time-dependent PDEs in general) should not forego the implementation of high-order methods in their efforts to parallelize their solvers. Further quantitative studies of the relative performance of CHASTE and Nektar++ are ongoing.

References

[1] Tung L 1978 *A bi-domain model for describing ischemic myocardial D-C potentials* Ph.D. thesis MIT Department of Electrical Engineering and Computer Science
[2] Linge S, Sundnes J, Hanslien M, Lines G and Tveito A 2009 *Philos. Trans. R. Soc. A* **367** 1931–1950
[3] Kerckhoffs R C P, SN H, Usyk T and McCulloch A 2006 *Proc. of the IEEE* **94** 769–783
[4] Niederer *et al* 2011 *Philos. Trans. R. Soc. A* **369** 4331–4351
[5] 2012 CHASTE `http://www.cs.ox.ac.uk/chaste/` [Online; accessed 30-May-2012]
[6] Bernabeu M O, Bordas R, Pathmanathan P, Pitt-Francis J, Cooper J, Garny A, Gavaghan D J, Rodriguez B, Southern J A and Whiteley J P 2009 *Philos. Trans. R. Soc. A* **367** 1907–1930
[7] Nektar 2012 Reference manual to Nektar++ library `http://www.nektar.info/wiki/3.2/Reference` [Online; accessed 30-May-2012]
[8] Vos P E J, Sherwin S J and Kirby M R 2010 *J. Compt. Phys.* **229** 5161–5181
[9] Vos P E J, Eskilsson C, Bolis A, Chun S, Kirby R M and Sherwin S J 2011 *Int. J. Compt. Fluid. Dyn.* **25** 107–125
[10] Burrage K and Butcher J 1980 *BIT.* **20** 185–203
[11] Ascher U M, Ruuth S J and Spiteri R J 1997 *App. Num. Math.* **25** 151–167
[12] Ethier M and Bourgault Y 2008 *SIAM J. Num. Analy.* **46** 2443–2468
[13] Ascher U M, Ruuth S J and Wetton B T R 1995 *SIAM J. Num. Analy.* **32** 797–823
[14] FitzHugh R 1961 *Biophys. J.* **1** 445–466
[15] Sundnes J, Lines G T, Cai X, Nielsen B F, Mardal K A and Tveito A 2006 *Computing the electrical activity in the heart* (Berlin: Springer-Verlag)

High Performance Computing Symposium 2012 (HPCS2012) IOP Publishing
Journal of Physics: Conference Series **385** (2012) 012010 doi:10.1088/1742-6596/385/1/012010

Towards the contention aware scheduling in HPC cluster environment

Sergey Blagodurov
Simon Fraser University
E-mail: sergey_blagodurov@sfu.ca

Alexandra Fedorova
Simon Fraser University
E-mail: alexandra_fedorova@sfu.ca

Abstract.
Contention for shared resources in High-Performance Computing (HPC) clusters occurs when jobs are concurrently executing on the same multicore node (there is a contention for shared caches, memory buses, memory controllers and memory domains). The shared resource contention incurs severe degradation to workload performance and stability and hence must be addressed. The state-of-the-art HPC clusters, however, are not contention-aware. The goal of this work is the design, implementation and evaluation of a virtualized HPC cluster framework that is contention aware.

1. Introduction

High Performance Computing applications pose great demands on the clusters that host them. Highly parallel in nature, they are able to fully utilize computational cores allocated to them on the cluster multicore nodes. Although cluster schedulers typically put each process (rank) on its own CPU core, the processes collocated on the same node are not isolated from each other. On a multicore server, cores share parts of the memory hierarchy, such as caches, system request queues, memory controllers and memory domains [9, 10, 17, 7]. The job processes compete for those shared memory resources and suffer performance penalty as a result.

In this paper we present our solution to the problem of shared resource contention in HPC. Our contention-aware HPC framework enables cluster administrators to detect and mitigate contention effects on cluster compute nodes. We address the problem both on node level via scheduling on each cluster node and on cluster level via live migration of the MPI cluster workload from the contended servers. Mitigating shared resource contention results in faster runtimes for the cluster workload.

This paper provides the following contributions:

1) We summarize our experiments performed on a number of industry level multicore servers and identify those hardware that is prone to the shared resource contention from the cluster workload.

2) Not all of the cluster workloads are contention-sensitive. Our framework allows to detect those that suffer from the shared resource contention on-the-fly with a robust, non-intrusive profiling method.

3) Our proposed framework addresses shared resource contention in two ways:

a. Our contention-aware scheduling algorithm runs on every compute node. It periodically rebinds job processes to cores and migrates process memory between server NUMA domains to reduce shared resource contention on the node. This is done automatically on-the-fly. No involvement from the cluster administrators is necessary.

b. Higher performance benefits are possible if the load can be spread away from the contended node. Our framework provides cluster administrators with such opportunity. The framework reports those nodes that experience high shared resource contention and allows live migration of process ranks across the compute nodes in the cluster. This is done through a light-weight OS-level virtualization installed in the framework. We provide further details below.

The rest of this paper is organized as follows: Section 2 defines the problem scope and answers in what conditions shared resource contention is the most severe. Section 3 introduces our cluster framework and provides summary of contributions made within the project. Section 4 presents experimental results. Section 5 concludes the paper with the discussion of possible ways the developed framework can be used and the description of our future steps.

2. Problem Scope

Shared resource contention appears when the processes compete for the access to the memory resources of a multicore server, such as last-level caches (LLC), system request queues, memory controllers and memory domains. When these resources are a bottleneck, the contention results in performance degradation for the server workload. To identify the hardware that is prone to contention, we have conducted extensive experimentation on several industry level multicore servers manufactured by Dell, HP and IBM with the processors from both AMD and Intel. All servers were running Linux 2.6.32. We used applications from SPEC MPI2007 as our workload. MPI2007 is a benchmark suite for evaluating MPI-parallel, floating point, compute intensive performance across a wide range of cluster and SMP hardware. These benchmarks pose high requirements for the memory capacity of the compute nodes: typically 1 GB per rank for the medium benchmark subset and 2 GB per rank for the large subset.

We summarize our findings as follows:

1) The servers built with Intel processors based on Core microarchitecture (Xeon X5365, Xeon E5320) and all the AMD Opterons we tested (2350 Barcelona, 8356 Barcelona, 2435 Istanbul, 8435 Istanbul) are prone to the shared resource contention. In our previous work, we have identified Front Side Bus and memory controllers to be the main bottlenecks, although all levels of the memory hierarchy are contributing to the contention overhead [10, 9, 17, 7]. Due to space limitations, we only present experimental results for Opteron servers in Section 4.

Servers built with Intel Nehalem microarchitecture (Xeon X5570, Xeon X5650 were tested) do not exhibit a significant shared resource contention. Memory hierarchy of these machines tends to perform better under memory intensive CPU-bound load. There is, however, one exception. We detected high performance degradation when HyperThreading feature is enabled. Although previous studies have reported that some HPC workloads can benefit from HyperThreading [1], we do not believe that it is widely used in HPC these days. Hence, we omit HyperThreading results from this paper (the interested reader can access them at [4]).

2) Performance degradation that applications experience from the shared resource contention is not uniform. The important question is then what metric to use in order to characterize the processes as sensitive to resource contention within the memory hierarchy of a compute node?

Consider a workload of memory-intensive applications, i.e., applications that are characterized by a high rate of requests to main memory. Following the terminology adopted in an earlier study [16] we will refer to these applications as *devils*. Applications with a low rate of memory requests are referred to as *turtles*. Our methodology from the earlier work [9, 10, 17] allowed us to identify the *LLC miss rate*, which is defined to include all requests issued by LLC to main memory including pre-fetching, as one of the most accurate predictors of the degree to which applications will suffer when co-scheduled in the memory hierarchy of a multicore node. We consider the process a devil if it has more than 30

LLC misses per 10,000 retired instructions. Otherwise, the process is a turtle. We further divide devils into two subcategories: *regular devils* and *semi-devils*. Regular devils have a miss rate that exceeds 45 misses per 10,000 instructions. Semi-devils have between 30 and 45 misses per 10,000 instructions. We explain why such classification is necessary in Section 4. We obtain LLC missrate online using hardware performance counters available on all modern processor models. Profiling with performance counters is robust and does not impose slowdown to the workload. We have also determined that most of the programs from SPEC MPI 2007 are devils and prone to shared resource contention (see Section 4 for details).

3. Contention-aware HPC framework

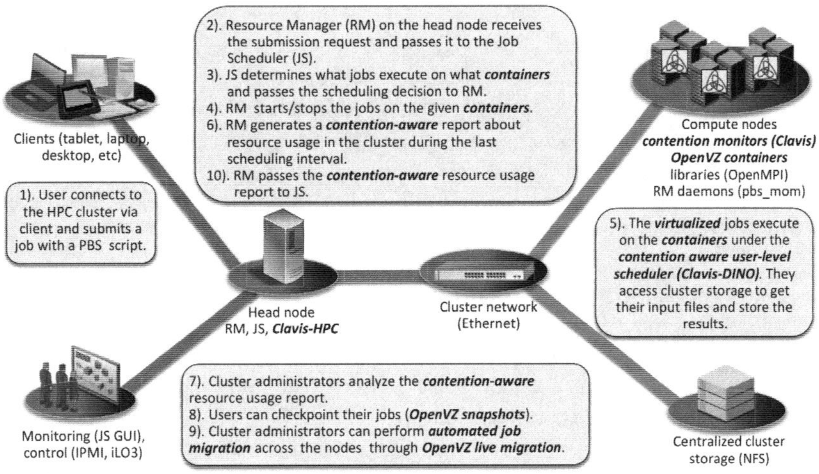

Figure 1: HPC cluster setting with contention-awareness.

The point of this work is to augment a typical HPC cluster installation with contention awareness, which it is currently lacking. Figure 1 depicts our contention-aware HPC framework and a job management cycle in it. If there are job submission requests posted in the last scheduling interval, the framework goes through steps 1–10. Otherwise, only steps 3–10 are being executed. For the most part, the management cycle is self-explanatory. We will clarify the remaining parts below. One thing we would like to highlight at that point is that there is a *two level scheduling* typically present in an HPC cluster: (a) cluster-wide scheduling on the head node (step 3) and (b) the OS-wide scheduling performed within each compute node (step 5). The first level decides what job processes to put on what nodes in a cluster, the second level then deals with scheduling of job processes assigned to a given node within that node. The first level scheduling is done by the job scheduler, the second level is usually done by the OS kernel running on the node.

The contention-related modifications are highlighted in italic. The list of changes include:

1) A virtual framework (OpenVZ) on each cluster node that allows installing Resource Manager monitoring tools (pbs_moms) and software libraries (OpenMPI, etc) inside the virtual machines (OpenVZ containers) instead of on the nodes themselves. This allows running jobs inside the containers as opposed to baremetal. The actual compute nodes are thus hidden from resource manager (RM) and cluster job scheduler (JS), they now treat virtual containers as nodes when performing cluster-wide scheduling (steps 3 and 4) or when checkpointing jobs for fault tolerance (step 8).

2) Online detection of the contention-sensitive jobs as they run on the compute nodes with Clavis user level daemon. Using a contention-aware scheduling algorithm called DINO within each node as opposed to the default OS scheduler to address shared resource contention locally (step 5).

3) Using the data obtained from Clavis, the framework periodically generates a report about shared resource contention in the cluster. It then shares the report with the job scheduler and cluster administrators upon request (steps 6, 7 and 10).

4) Using the information in the report, cluster administrators can then live migrate OpenVZ containers running memory-intensive MPI processes to the idle nodes on-the-fly, thus reducing job execution time. The framework allows to do it independently of the job scheduler, without affecting its job queues (step 9).

The instructions and scripts on deploying the framework are available at [4]. We next describe our modifications in more details.

3.1. HPC framework virtualization

Virtualization is a technique that allows running several OS instances simultaneously within a single hardware node. It is wildly adopted in the cloud, and some projects suggested to use the cloud computing resources like Amazon to host HPC clusters for educational purposes, e.g. StarHPC from MIT [13]. However, so far virtualization did not gain much popularity in HPC. Supercomputers are still largely scheduling jobs on their compute nodes baremetal. One of the main reasons is that virtualization is not free, there is an overhead of using it. The cluster workloads pose great demands on CPU, memory, network and IO subsystems of the supercomputer. And while CPU virtualization is not a problem these days due to Hardware Assisted Virtualization [2] implemented in most modern processor models, the overhead of virtualizing memory, network and IO can be substantial [14, 11, 15].

There exist many different virtualization solutions, each of which utilizes its own approach to virtualization. A good candidate for an HPC-based virtualization must provide near native memory, network and IO performance for the cluster MPI workloads. The Open Source OS-level virtualization solution called OpenVZ does offer such capabilities. Unlike other solutions (KVM, VMWare, Xen to name a few) that can run multiple operating systems at once, OpenVZ uses a single patched Linux kernel that is running both on the host and in the guests. It uses a common file system so each VM is just a directory of files that is isolated using chroot and special Linux kernel modules. However, because it does not have the overhead of a full hypervisor, it is very fast and efficient. Different studies have shown an average performance overhead of 1–3% for virtualizing OpenVZ as compared to using a standalone server [3, 14].

Memory management in OpenVZ highly favors contention-aware scheduling on NUMA multicore nodes. In other virtualization solutions memory management is usually performed by the guest and host OSes independently. It is generally very hard to say which pages of a total VM memory footprint belong to a particular MPI rank that is executing inside the guest. The memory in OpenVZ, on the other hand, is managed by the host kernel. As a result, the migration of the guest memory between different NUMA domains is as easy as migrating memory of the workload executing baremetal.

We found the following considerations to be important when using OpenVZ for the HPC-based virtualization:

1) OpenVZ does support live migration. Live migration allows moving a full process environment (processor context, memory, sockets, etc) between the hardware nodes on-the-fly. This is different from what is used in HPC nowadays, namely checkpointing a job, terminating it and restoring on different set of nodes. If done correctly, live migration does not terminate the open network connections of the job.

OpenVZ uses the approach to live migration that can be described as "suspend – move – resume", in which the VM is being frozen and its whole state is saved to a file on disk. This file is then copied to another machine and the VM can then be unfrozen (restored) there. This approach is different from incremental memory migration used in KVM and Xen in which the source VM continues running while certain pages are moved across the network to the new machine. To ensure consistency, pages modified during this process must be re-sent. However, if the memory footprint changes too rapidly, the incremental approach may simply hang. That is why the "suspend – move – resume" approach may be more reliable, although it could take longer to complete. Although we only use Open Source tools within this project, we note that a proprietary virtualization solution called Virtuozzo which is built on top of OpenVZ does support incremental memory migration [5].

2) For the live migration to be successful, the VM memory footprint should be locked pages free. That assumes the PBS and MPI libraries installed inside the VM as well as the cluster workload running on it should be compiled without locked pages support.

3) OpenVZ does not offer an InfiniBand support, so it is currently applicable only to the clusters with Ethernet interconnect. This is a disadvantage relative to other virtualization solutions that do support InfiniBand, KVM being one example [12]. However, developers may include IB support in future releases of OpenVZ [14].

In order to run cluster jobs inside the containers as opposed to baremetal, we install RM monitoring tools (pbs_moms) and software libraries (OpenMPI, etc) inside the OpenVZ virtual containers instead of on the nodes themselves. The actual compute nodes are now hidden from RM and JS, they now treat containers as nodes when performing cluster-wide scheduling. That allows to seamlessly, without interfering with JS or RM work, move the workload across the nodes if necessary. When deciding how many containers to create per each hardware node, we must take into account the number of physical cores, memory and the disk space available. Although most of the kernel modules are shared between containers on the node, each container creates its own Linux file structure on the disk. For our experimental machines, we found that the two containers per node is a reasonable value: it allows to mitigate the contention effects while not consuming much of a disk space or network traffic during migrations.

3.2. The node-level scheduling

Previous studies have demonstrated that the contention effects can incur severe degradation to the CPU-intensive cluster workloads [9, 6]. At the same time, the default Linux kernel scheduler that usually assigns ranks to cores within each compute node is not contention aware. In order to address this gap, we created Clavis. Clavis is a user level daemon written in C that is designed to implement various scheduling strategies under Linux running on multicore and NUMA machines. It is released under Academic Free License (AFL) and its source is available for download [8].

Clavis is started on each physical server when it is booted. It then proceeds as follows:

1) It automatically detects the OpenVZ containers present on the system and compute bound MPI ranks running in each of them. It then monitors the LLC missrate of the thus detected processes and classifies them as either devils or turtles.

3) It schedules the detected ranks using its contention-aware scheduling algorithm. For the NUMA multicore servers, we use the algorithm called Distributed Intensity NUMA Online (DINO) [9]. DINO separates devils in the memory hierarchy of the multicore system as far from each other as possible, thus reducing contention for memory hierarchy of the node. It uses LLC missrate metric to detect the memory intensiveness of the applications (whether its a devil or a turtle). When DINO moves the process away from its memory on the NUMA memory socket, it later pulls the memory close to the process to preserve memory locality.

4) Finally, it creates a concise profile of the workload in each container and saves it on the cluster file server. The report includes information about CPU utilization, allocated memory size, traffic and contention class of the MPI ranks running inside the container.

Clavis is able to improve workload performance on the node level in the following situations:

1) When all the ranks executing on the node are memory intensive (devils). The performance improvements in this case are modest, because, no matter how DINO schedules the ranks across the cores within a node, the shared resource contention still prevails. However, it is still possible to provide improvement relative to Linux Default scheduler in this case due to better memory locality achieved under DINO. We provide the experimental results in Section 4.

2) When devils are running alongside with turtles on the compute node. We studied this situation extensively before [9, 10].

3) When ranks are restored on the new compute node after live migration. Although live migration mechanism is able to fully restore process working environment on the new node, the distribution of the process' memory across NUMA domains after migration is completely random (hypervisor just restores the pages on the first NUMA domain available). Hence, proper scheduling of processes and their memory by DINO is able to rip big benefits in comparison with Default Linux scheduler that does not perform memory migration at all (Section 4).

3.3. Contention-aware resource usage reports

We now need the ability to aggregate the resource profiles collected by Clavis daemons on various compute nodes into contention-aware resource usage reports for HPC cluster as a whole. This is done by the RM. We use a popular Open Source RM called Torque. Its pbs_mom and pbs_server tools were modified within this work to periodically perform the following:

In each scheduling interval, Torque generates a list of active containers by parsing the *pbsnodes* command output. It reads the Clavis resource profiles for each container from the cluster storage. Torque determines which MPI ranks belong to which cluster jobs. The information aggregated per job in Torque is then supplied to the cluster scheduler via standard communication means.

3.4. The cluster-wide scheduling with Clavis-HPC

The task of a cluster scheduler is to maintain job queues and decide which nodes (containers in our implementation) will host the job ranks during execution. The scheduler does not consider contention effects in its scheduling decisions. It is also not able to migrate the load across the cluster on-the-fly, away from the contended nodes. To demonstrate the potential of our framework, we created Clavis-HPC, a user level cluster-wide load balancer written in Python. It runs in parallel with Maui on the head node and allows cluster administrators to seamlessly live migrate containers, along with the MPI ranks inside them, across the cluster. Clavis-HPC can be used to reduce shared resource contention as follows:

1) The cluster administrator uses Maui command-line tools to view the contention-aware resource usage report. The report includes information about which containers host contention-intensive workload (we call such containers *devilish*) and on what nodes these containers are located.

2) If there are compute nodes that host 2 devilish containers, it is worthwhile to migrate one of them away from this contended node, thus reducing the shared resource contention on it. There can be two cases: (a) the devilish container is migrated to the idle node and (b) the devilish container is swapped with a container that host turtles on some other node.

These load balancing decisions do not interrupt the active cluster workload or get in the way of the Maui queue management. Section 4 further describes the results of our scheduling.

4. Evaluation

We use the following facilities for testing.

Dell_Opteron: a small experimental 48-core cluster constructed from 3 servers. The servers have the following hardware configuration:

Two Dell-Poweredge-R805 (AMD Opteron 2435 Istanbul) servers have twelve cores placed on two chips. Each chip has a 6MB cache shared by its six cores. It is a NUMA system: each CPU has an associated 16 GB memory block (32 GB in total). Each server had a single 70 GB SCSI hard drive. We denote these machines as nodes X and Y.

One Dell-Poweredge-R905 (AMD Opteron 8435 Istanbul) server has 24 cores placed on four chips. Each chip has a 5 MB L3 cache shared by its six cores. Each core also has a private unified L2 cache and private L1 instruction and data caches. It is a NUMA system: each CPU has an associated 4 GB memory block, (16 GB in total). The server was configured with a single 76 GB SCSI hard drive. We treat this machine as two 12 core nodes with 2 NUMA domains each. We denote these nodes V and Z.

The servers were configured with Linux Gentoo 2.6.32 release 9. We use Maui as our cluster scheduler, Torque as a resource manager and SPEC MPI2007 as our workload. We run all the MPI jobs with 24 ranks in total. 1GbE network is used as an interconnect between all the nodes. In case of the nodes V and Z, the traffic passes through the same 1GE interface to get to the other nodes. Sharing the same interface may be the bottleneck. To reduce its influence, we use nodes X and Y for our 12 rank per node runs. We only use nodes X, Y, V, Z for 6 ranks per node runs, thus minimizing the traffic via the shared interface (see below).

First of all, we have decided to perform the experiments to see if our DINO contention-aware algorithm provides any benefits to the MPI cluster workloads when scheduling on the node level. Series A on Figure 2 shows the average performance improvement for devils and turtles across different runs on Dell_Opteron cluster. As can be seen, DINO outperforms Linux Default for most of the applications tested.

Next, we wanted to demonstrate the benefits from the contention-aware cluster-wide scheduling. Series B shows that, for all the devils, there is a benefit in scheduling a job with 6 ranks per node (and taking twice as many nodes) rather than scheduling it with 12 ranks per node. The challenge here is to detect which ranks are devils and then spread only those apart. Series C–F shows the performance improvement of a solution where the job class is detected online and devils are spread across the nodes with OpenVZ live migration feature. We make the following conclusions from these results:

1) OpenVZ containers incur little overhead on its own (Series C). There can be performance degradation after migration though, if contention-unaware algorithm (like Linux Default) is used on the cluster nodes (Series E). Our contention-aware DINO algorithm (Series F), on the other hand, is able to provide solution, which is close to the best separated one of Series B (when ranks are started on different nodes from the start and no migration occurred).

2) Separating devil processes across nodes is beneficial for the overall performance of an MPI job (Series F). The influence of such separation is twofold: it relieves pressure on shared resources within a node, but it also increases pressure on the network. Series F results state that, even for a slow 1GbE connection and migration overhead included, performance benefit of reducing shared resource contention outweighs the interconnect slowdown for all devils. Those jobs that cannot benefit from separation (turtles) or those for whom migration penalty outweighs performance benefits (semi-devils) can be dynamically detected by the framework and used for load balancing (to plug the idle slots on partially loaded nodes). It is generally very hard to precisely measure what is the impact of each of the two conflicting factors on the overall performance of a job, since they interact in complex ways. Series D provides an estimate of the sensitivity of our workloads to exchanging data via the 1GbE interconnect. We obtained it by increasing the number of jobs running on the node: instead of 12 ranks of a single job, the ranks of two jobs were present on the node (6 ranks of each job). Both jobs were different instances of the same program. Such experimental configuration keeps the shared resource contention on the same level (12 ranks per node), but it increases the pressure on the network. Series D shows that most of the programs we considered do not take big hit when using the network and so the performance benefit of Series F is roughly due to the relief in shared resource contention. We do work on a metric that characterizes the interconnect sensitivity of a job, just like a LLC miss rate characterizes its sensitivity to shared resource contention. However, due to space limitations, we leave this topic outside the scope of the paper.

5. Conclusion

In this paper we demonstrated how the shared resource contention can be addressed when creating a scheduling framework for a scientific supercomputer. The proposed solution is comprised of the Open Source software that includes the original code and patches to the widely-used tools in the field. The solution (a) allows an online identification of the contention-sensitive jobs and (b) provides a way to make and enforce contention-aware scheduling decisions both on cluster level and within each multicore node.

We are currently working on improving our algorithms to simultaneously satisfy several conflicting scheduling goals (reduce contention, reduce power, reduce communication overhead) within the

Journal of Physics: Conference Series **385** (2012) 012010 doi:10.1088/1742-6596/385/1/012010

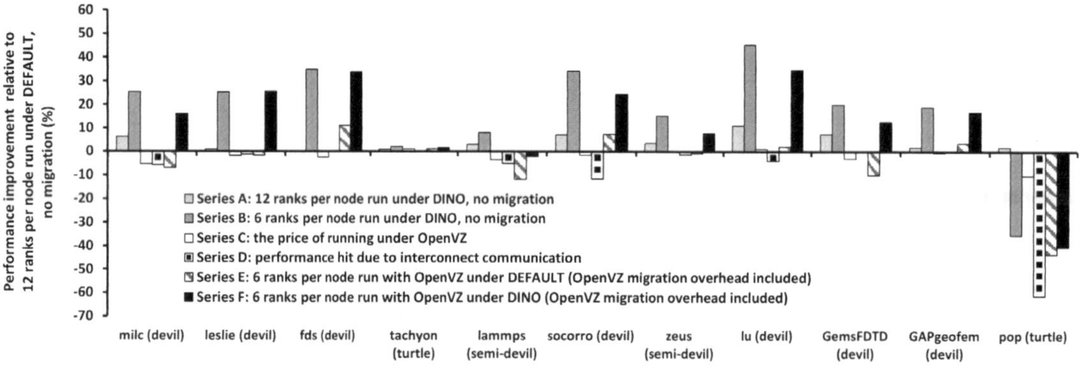

Figure 2: Results of the contention-aware experiments on Dell_Opteron.

described cluster environment simultaneously.

6. Acknowledgment

This material is based upon work supported in part by the National Science Foundation under Grant No. 0910812 to Indiana University for "FutureGrid: An Experimental, High-Performance Grid Test-bed." Partners in the FutureGrid project include U. Chicago, U. Florida, San Diego Supercomputer Center - UC San Diego, U. Southern California, U. Texas at Austin, U. Tennessee at Knoxville, U. of Virginia, Purdue U., and T-U. Dresden.

[1] A Study of Hyper-Threading in High-Performance Computing Clusters. *[Online] Available: http://www.dell.com/content/topics/global.aspx/power/en/ps4q02_leng.*
[2] Hardware Assisted Virtualization. *[Online] Available: http://en.wikipedia.org/wiki/Hardware-assisted_virtualization.*
[3] Open VirtualiZation. *[Online] Available: http://en.wikipedia.org/wiki/OpenVZ.*
[4] Optimizing Shared Resource Contention in HPC Clusters (project webpage). *[Online] Available: http://hpc-sched.cs.sfu.ca/.*
[5] Zero-downtime migration. *[Online] Available: http://download.swsoft.com/virtuozzo/virtuozzo4.0/ docs/en/lin/VzLinux-UG/14373.htm.*
[6] Blagodurov, S., and Fedorova, A. In search for contention-descriptive metrics in HPC cluster environment. ICPE.
[7] Blagodurov, S., and Fedorova, A. User-level scheduling on NUMA multicore systems under Linux. In *Linux Symposium* (2011).
[8] Blagodurov, S., and Fedorova, A. User-level scheduling on NUMA multicore systems under Linux. In *Linux Symposium* (2011).
[9] Blagodurov, S., Zhuravlev, S., Dashti, M., and Fedorova, A. A Case for NUMA-Aware Contention Management on Multicore Systems. In *USENIX ATC* (2011).
[10] Blagodurov, S., Zhuravlev, S., and Fedorova, A. Contention-aware scheduling on multicore systems. *ACM Trans. Comput. Syst. 28* (December 2010), 8:1–8:45.
[11] El-Khamra, Y., Kim, H., Jha, S., and Parashar, M. Exploring the Performance Fluctuations of HPC Workloads on Clouds. CLOUDCOM.
[12] Hillenbrand, M., Mauch, V., Stoess, J., Miller, K., and Bellosa, F. Virtual infiniband clusters for hpc clouds. CloudCP '12.
[13] Ivica, C., Riley, J., and Shubert, C. StarHPC – Teaching parallel programming within elastic compute cloud.
[14] Regola, N., and Ducom, J.-C. Recommendations for Virtualization Technologies in High Performance Computing. CLOUDCOM.
[15] Tikotekar, A., Vallée, G., Naughton, T., Ong, H., Engelmann, C., and Scott, S. L. Euro-par 2008 workshops - parallel processing. 2009, ch. An Analysis of HPC Benchmarks in Virtual Machine Environments.
[16] Xie, Y., and Loh, G. Dynamic Classification of Program Memory Behaviors in CMPs. In *CMP-MSI* (2008).
[17] Zhuravlev, S., Blagodurov, S., and Fedorova, A. Addressing Contention on Multicore Processors via Scheduling. In *ASPLOS* (2010).

High Performance Computing Symposium 2012 (HPCS2012) IOP Publishing
Journal of Physics: Conference Series **385** (2012) 012011 doi:10.1088/1742-6596/385/1/012011

A Hybrid Double-Layer Master-Slave Model For Multicore-Node Clusters

Gang Liu, Hartmut Schmider and Kenneth E. Edgecombe
High Performance Computing Virtual Laboratory (HPCVL),
Queen's University, 115-993 Princess St., Kingston, Ontario, K7L 1H3, Canada
E-mail: gang.liu@queensu.ca

Abstract. The Double-Layer Master-Slave Model (DMSM) is a suitable hybrid model for executing a workload that consists of multiple independent tasks of varying length on a cluster consisting of multicore nodes. In this model, groups of individual tasks are first deployed to the cluster nodes through an MPI based Master-Slave model. Then, each group is processed by multiple threads on the node through an OpenMP based All-Slave approach. The lack of thread safety of most MPI libraries has to be addressed by a judicious use of OpenMP critical regions and locks.

The HPCVL DMSM Library implements this model in Fortran and C. It requires a minimum of user input to set up the framework for the model and to define the individual tasks. Optionally, it supports the dynamic distribution of task-related data and the collection of results at runtime. This library is freely available as source code. Here, we outline the working principles of the library and on a few examples demonstrate its capability to efficiently distribute a workload on a distributed-memory cluster with shared-memory nodes.

1. Introduction

Modern clusters usually consist of nodes with multiple CPUs or cores. In addition, recent core architectures support multiple threads being concurrently executed on a single core (hyper-threading). The standard approach to exploit parallelism on a cluster is by use of the Message-Passing Interface MPI. This requires the communication between independent "heavy-weight" processes that are deployed across the nodes in a distributed-memory fashion. While it is possible to run multiple independent MPI processes on a single multicore cluster node, it may not be the most efficient use of resources. This is particularly true for multithreaded cores which exhibit a high degree of resource sharing between threads, and are therefore prone to run into bottlenecks, particularly connected to communication resources.

A more appropriate approach to use multicore and multi-threaded nodes is through the use of thread libraries or compiler directives such as OpenMP. The resulting multithreaded programs do not require communication resources and use shared memory for information exchange between "light-weight" processes instead. As a result, resource-bottlenecks are much less likely. Unfortunately, due to their dependence on the shared-memory architecture of the computer, such programs cannot be deployed across a cluster.

It therefore appears that the ideal approach for programming for a multicore, multi-threaded cluster, is a hybrid one: combining MPI to deploy processes across the cluster to the individual nodes, with OpenMP parallelization of each process to make efficient use of the multi-thread structure of the hardware.

The main complicating issue with this approach is the lack of thread safety of most MPI libraries. In practice this means that communication resources cannot be accessed by more than one thread per process at a time, because internal communication buffers and other variables are not protected against race conditions. As a result the programmer must ensure that MPI is consistently seeing only one thread per rank and that all information that is communicated is made available to all threads at a time when they need it.

In this paper, we introduce a library implementing a simple model that is applicable in cases when independent tasks have to be performed and no intermittent communication is required, the "Double-Layer Master Slave Model" (DMSM). This model combines a standard Master-Slave approach for the workload distribution among the nodes of a cluster, with a local "All-Slaves" scheme on each of the nodes to work through a group of tasks.

Similar models have been discussed previously under various names. A multi-level Master-Slave model was presented by Chen, Lim, and Warsi to overcome communication overheads on the master [1]. Leopold and Süß used a multi-layered hybrid approach to schedule a Global-Water Simulation Program on a heterogeneous cluster [2, 3]. The scalability of "Bag-of-Tasks" applications was the subject of a systematic study by da Silva and Senger [4]. Another hybrid MPI/OpenMP approach was taken Wu, Lai, and Chiu [5] to address heterogeneities on multi-core clusters.

The emphasis of this paper is on the implementation of the DMSM model in the form of a freely available library.

2. The Double-Layer Master-Slave Model

The MPI implementation of the classic Master-Slave Model can readily be expanded to a hybrid model if tasks are combined into "workload packages" and distributed across the nodes. Each of the packages is then processed on a node via an OpenMP All-Slaves Model.

(i) On a cluster, one node serves as Master Node, and thread 0 is dedicated as Master. The other nodes are Slave Nodes. Communication is done via MPI calls among nodes.

(ii) In the beginning, each Slave signals to the Master that it is idle, and the Master sends out workload packages that are handled internally by the Slave nodes.

(iii) Any Slave Node works through a package without any further communication among nodes. Once the workload is executed, another package is requested. This happens until a stop signal is received.

(iv) The Master continues to serve requests until all work is distributed. Then a stop signal is passed to the Slaves. The Master insures that no package is distributed more than once or skipped.

In the above Master-Slave scheme the workload execution on the nodes (Step iii) is in itself done through OpenMP:

(i) Once a new workload package is received on a Slave node, a task counter is reset, and multiple OpenMP threads begin to work on the package.

(ii) Each thread assigns itself to a given task in the package and moves the task counter forward to avoid double execution or skipping.

(iii) The task counter is protected by placing updates into a critical region or by use of locks.

(iv) Once all tasks in a package are executed or at least assigned, a single thread communicates with the Master for a new package.

Since on a given node, memory is shared and any thread can access the task counter, there is no need for a master thread. The resulting model can adequately be described as an All-Slave Model. The details give rise to different variations, but it is essential that only one thread at a time communicates with the master node, as MPI libraries are usually not thread-safe.

3. The HPCVL DMSM Library

This library was developed by Gang Liu since 2009 to enable the deployment of a large number of independent computing tasks with unpredictable execution times across a cluster with multithreaded multicore nodes. The library is implemented in Fortran 90 and C. The current version is 3.1. The library is available as source code, free of charge [6].

3.1. Main interface and model variations

The core part of the library is an interface that allows the user to specify parameters for the DMSM model:

```
SUBROUTINE DMSM_ALL(        &  ! full DMSM run, including initialization and finalization
THREADS_PER_PROCESS,        &  ! number of OpenMP threads per MPI process (node)
JOB_DISTRIBUTION_PLAN,      &  ! specifies variation of the model (see Table 1)
TOTAL_JOBS,                 &  ! total number of tasks to be performed
NUM_OF_JOBS_PER_GROUP,      &  ! number of tasks per package
DO_THE_JOB,                 &  ! routine performing one job (required)
JOB_GROUP_PREPARATION,      &  ! routine communicating data for package (optional)
RESULT_COLLECTION,          &  ! routine collecting results from nodes (optional)
COLLECTION_ENABLED)            ! controls results collection (optional)
```

The only mandatory user routine is `DO_THE_JOB` with one argument, the task number. This implements the execution of a single task. The other routines are optional and will be discussed in section 3.2.

The parameter `JOB_DISTRIBUTION_PLAN` determines what variation of the DMSM is executed: The first column indicates the argument that is passed to `DMSM_ALL`. The second

Table 1. Variations of the Model

Arg	Other threads on Master	Slave Communication
11	none	0 from serial region
12	none	0 from parallel region
13	none	any from parallel region
21	pre-allocated	0 from serial region
22	pre-allocated	0 from parallel region
23	pre-allocated	any from parallel region
31	dynamic	0 from serial region
32	dynamic	0 from parallel region
33	dynamic	any from parallel region

column describes the workload allocation for threads on the master node other than the master thread. The third column describes which thread on the slave nodes is communicating with the master node. Model `11` is the least flexible: only one thread on the master node is active, and OpenMP parallelism on the slave nodes is completely separate from MPI communication, meaning that whenever a thread runs out of work, a new package will only be sent to the node when all other threads are finished. Model 33 is the most flexible: all threads on the master node work through packages dynamically, and on the slave node, new packages are requested and sent as soon as one thread runs out of work.

3.2. Group preparation, result collection and locks

In many cases, initial data have to be distributed to the nodes before work on each task package can commence. For this the user may specify another routine `JOB_GROUP_PREPARATION` which is called whenever a new package is requested. To address a race condition that exists when new data are distributed while the previous set is still in use, a pair of routines `DMSM_WAIT_FOR_INITIAL_LOCKS` and `DMSM_UNSET_AN_INITIAL_LOCK` are supplied by the library. The locks that are used by these routines are not visible to the user.

In other cases, the results from the individual jobs can be collected to the Master node, either dynamically while executing the DMSM or at the end when all jobs are completed. For this purpose, the user can specify a third routine `RESULT_COLLECTION`. In addition, an integer parameter `COLLECTION_ENABLED` specifies whether dynamic collection is desired or not. If the collection happens every time a new task package is requested, another race condition may exist between the communication/reset of the temporary result data structure, and the accumulation of results that are still being generated. To address this, another pair of routines `DMSM_SET_NODE_RESULT_LOCK` and `DMSM_UNSET_NODE_RESULT_LOCK` are supplied. As before, the employed locks are not visible to the user.

These routines have to be used inside the user-supplied functions to protect data that are shared among local threads. The details of their usage are discussed in the User's Manual [7].

3.3. Other usages of the DMSM library

The DMSM library can also be used to run other model. For instance, a double-layer MPI model where the "outer layer" is a Master-Slave model, and the "inner layer" is a more communication intensive MPI parallelisation. Support for this in the library is implemented in the form of additional interfaces and MPI communicators. Likewise the parallelism on the nodes does not have to be in the form of an All-Slaves Model. For example, internal OpenMP parallelization may be implemented in the form of parallel loops.

4. Examples

In the following examples, we used the most flexible "Variation 33" throughout. This means that on the Master node, all threads other than the Master perform dynamically scheduled computations, while on the Slave node the first thread that runs out of work requests a new job package immediately. Job requests are protected through critical regions within the library.

4.1. Workload Balancing

To assess how efficiently the DMSM library balances a workload, we used a series of 851 partial optimizations of the H_2O_2 molecule with fixed angles and varying bond lengths. These were done through calls to the computational-chemistry software Gaussian'09 [8] from C-code using the DMSM library. As can be seen in Fig.1, the execution times for these optimizations differs greatly, between 50 and 200 seconds. We deployed the optimizations over 16 Sun T5140 nodes [9] with four serial runs ("threads") per node using the DMSM. The available multithreading within the Gaussian code was ignored. Fig.2 shows that the execution times per thread for the total run vary between 1000 and 1200 seconds, reducing the workload imbalance from 400% to 20%. This makes the DMSM a suitable means for throughput runs on clusters with high thread-count per node and low per-thread performance.

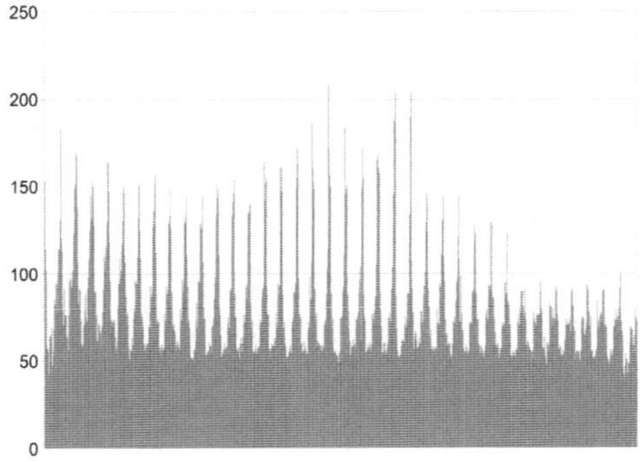

Figure 1. Total CPU time in seconds per job for 851 partial optimizations

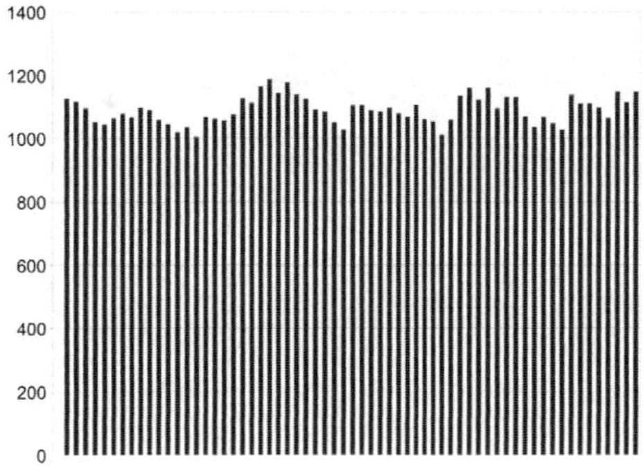

Figure 2. Total CPU time in sedconds per thread for 64 threads using DMSM

4.2. Single-node performance on a T4-2

To assess the performance of the DMSM on a single node, we used an Oracle T4-2 multithreaded server [9] that supports up to 128 threads. We ran the brute-force evaluation (by alternating series [10]) of 10,000 points of the Riemann Zeta function $\zeta(z)$ in the interval $2 \geq \mathrm{Re}(z) \geq 1.01$ and $1 \geq \mathrm{Im}(z) \geq 0.01$ in quadruple precision. This is a very slowly convergent series, and the required number of terms to reach a given accuracy depends strongly on the argument. This makes it suitable as a testcase. We call the number of MPI processes N, the number of OpenMP threads per MPI process n, the overall wallclock time T_w, and the accumulated job CPU time T_j. The total number of threads is $N_t = Nn$, and the number of slave processes (workers) is $N_s = N_t - 1$ for $N > 1$ and $N_s = N_t$ for $N = 1$.

- The theoretical workload efficiency (percentage of time spent doing work) increases with the number of threads: N_s/N_t.
- The real workload efficiency (T_j/T_w, Fig.3) is very close to the theoretical one except when many MPI processes are used.
- The scaling efficiency ($T_w[1,1]/(N_s T_w[N,n])$, Fig.4) is close to 100% for up to 16 threads (which is the number of cores), then declines but stays above 50%.

The 64-thread scaling performance of the T4-2 server is in the 60% range despite the 4-fold multithreading and the sharing of one floating-point processing unit among 8 threads. For the DMSM model this means that using 64 threads total on one node still improves performance, with a preference for small numbers of MPI processes. The similarity of the real workload efficiency to the theoretical one indicates that the DMSM is robust as long as not too many MPI processes are used.

4.3. Multi-node performance on a small T5140 Cluster

We used the T5140 cluster (see Gaussian experiment section 4.1) to test the DMSM library with evaluations of the Zeta function across multiple nodes. Running 64 evaluations per package on 64 threads (i.e. 4 per core), we scale up the number of nodes from 1 to 16, for a maximum number of 1024 threads.

The results indicate that a substantial loss of both workload efficiency (T_j/T_w, Fig.5) and scaling efficiency ($T_w[1]/(NT_w[N])$, Fig.6) occurs as the number of nodes is increased. This happens whether results are communicated back to the node or not, suggesting that it is the latency of the communication that hampers efficiency, not the size of the messages being passed.

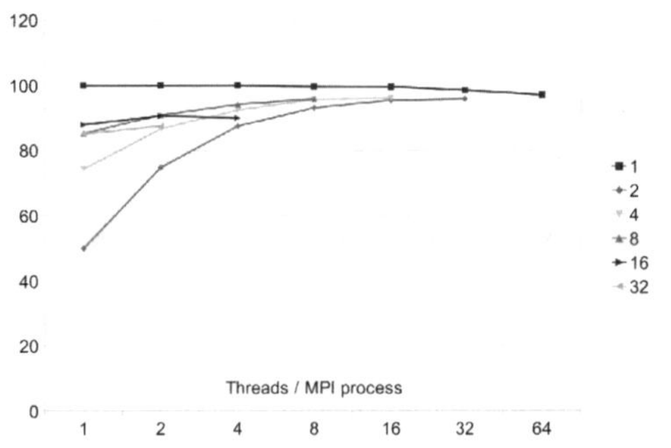

Figure 3. Workload efficiency for 10,000 functional evaluations of ζ, as a function of process number (MPI, abscissa) and threads per MPI process (OpenMP, legend).

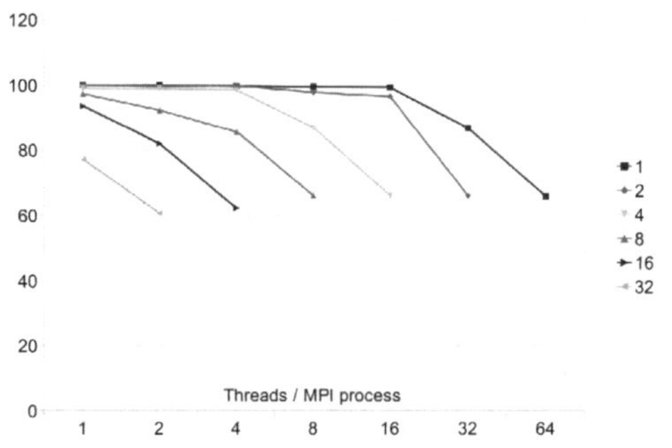

Figure 4. Scaling efficiency for 10,000 functional evaluations of ζ, as a function of process number (MPI, abscissa) and threads per MPI process (OpenMP, legend).

The fact that the decline in workload efficiency is very similar to the scaling one makes it likely that the former causes the latter. As the nodes are forced to communicate new job packages frequently, they spend less time working through jobs. Note that this issue is much less prominent in the single-node case, where communication is done through shared memory with very small latencies. Still, this example cannot be executed through a standard Master-Slave Model because the communication requirements of 1024 MPI processes would be prohibitive, nor through a simple All-Slaves model if no large enough shared-memory machine is available.

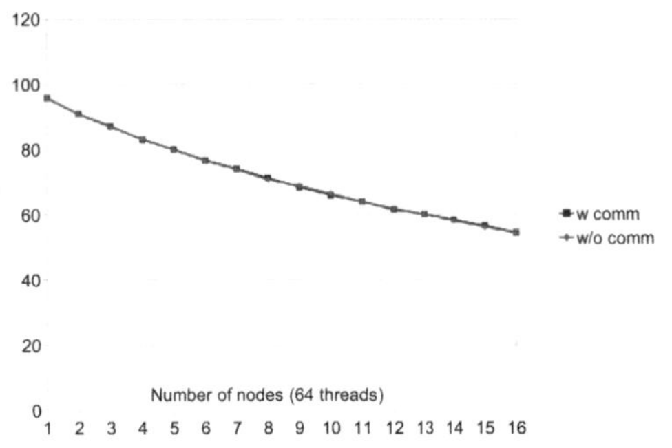

Figure 5. Workload efficiency for 10,000 functional evaluations of ζ, as a function of the number of nodes (MPI processes). The number of threads on the nodes is kept fixed at 16.

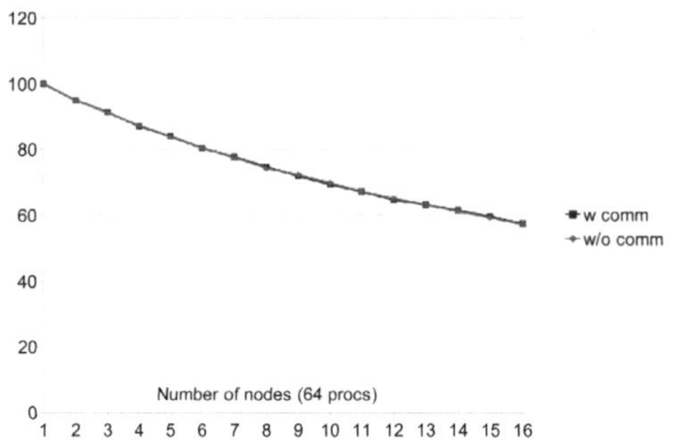

Figure 6. Scaling efficiency for 10,000 functional evaluations of ζ, as a function of the number of nodes (MPI processes). The number of threads on the nodes is kept fixed at 16. The single-node run is the reference.

5. Conclusions

The Double-Layer Master Slave Model (DMSM) is well suited to deploy a workload consisting of independent heterogeneous tasks on a cluster consisting of multicore nodes. The usage of a hybrid approach with an OpenMP model on the nodes and an MPI framework across the cluster addresses two issues that may occur in a simple Master-Slave (MS) model and a All-Slaves (AS) approach, respectively. Namely, in an MS model, multiple MPI processes on a multicore node tend to overwhelm the local shared resources, while an AS model can only be deployed on a single shared-memory node.

The HPCVL developed DMSM library implements this model in Fortran 90 and C, and is freely available as source code [6]. It provides a framework to run this model with a minimal required amount of user input but also allows some flexibility concerning the dynamic distribution of task-related data and the collection of results. Most importantly, it addresses the difficult issue of a lack of thread-safety of most MPI implementations by providing the proper lock functions and protecting tasks that are subject to race conditions.

References

[1] Chen, H.C. and Warsi, N.A., An efficient multilevel master-slave model for distributed parallel computation. **Boundary Element Technology XIII - Incorporating Computational Methods and Testing for Engineering Integrity**, Boundary Elements v. 2, p. 439-448, WIT Press, 1999.

[2] Leopold, C. and Süß, M., Observations on MPI-2 Support for Hybrid Master/Slave Applications in Dynamic and Heterogeneous Environments. **Proc. EuroPVM/MPI, LNCS**, v.4192, p. 285-292, Springer, 2006.

[3] Leopold, C., Süß, M. and Breitbart, J., Programming for Malleability with Hybrid MPI-2 and OpenMP: Experiences with a Simulation Program for Global Water Prognosis. **European Conference on Modelling and Simulation**, Bonn, p. 665-670, ECMS, 2006.

[4] da Silva, F.A.B. and Senger, H., Improving scalability of Bag-of-Tasks applications running on master-slave platforms, **Parallel Computing** v.35, p. 57-71, Elsevier, 2009.

[5] Wu, C.C., Lai, L.F. and Chiu, P.H., Parallel Loop Self-Scheduling for Heterogeneous Cluster Systems with Multi-Core Computers. **IEEE Asia-Pacific Services Computing Conference**, p. 251-256, IEEE Computer Society, 2008.

[6] The DMSM library can be downloaded at
http://www.hpcvl.org/sites/default/files/dmsm/DMSMcode.tar.
If the library is used in published work, the present paper should be cited as reference.

[7] Liu, G. and Schmider, H. User's Manual for the HPCVL DMSM Library,
http://www.hpcvl.org/sites/default/files/dmsm/DMSMmanual.pdf, HPCVL, 2011.

[8] Frisch, M.J. *et al.*, **Gaussian 09, Revision C.01**, Gaussian, Inc., Wallingford CT, 2010.

[9] The hardware used in the examples was a single Oracle T4-2 server with 256GB of memory
**http://www.oracle.com/us/products/servers-storage/servers/sparc-enterprise/\
t-series/sparc-t4-2**
and a 16-node cluster of Sun Sparc Enterprise T5140 servers with 32GB of memory per node
http://www.oracle.com/us/products/servers-storage/036000.pdf.

[10] Gourdon, X. and Sebah, P., Numerical evaluation of the Riemann Zeta-function.
http://numbers.computation.free.fr/Constants/Miscellaneous/zetaevaluations.pdf, 2003.

Journal of Physics: Conference Series **385** (2012) 012012 doi:10.1088/1742-6596/385/1/012012

GPU-Based Asynchronous Global Optimization with Particle Swarm

M P Wachowiak[1] **and A E Lambe Foster**
Department of Computer Science and Mathematics
Nipissing University, North Bay, ON Canada, P1B 8L7

E-mail: markw@nipissingu.ca

Abstract. The recent upsurge in research into general-purpose applications for graphics processing units (GPUs) has made low cost high-performance computing increasingly more accessible. Many global optimization algorithms that have previously benefited from parallel computation are now poised to take advantage of general-purpose GPU computing as well. In this paper, a global parallel asynchronous particle swarm optimization (PSO) approach is employed to solve three relatively complex, realistic parameter estimation problems in which each processor performs significant computation. Although PSO is readily parallelizable, memory bandwidth limitations with GPUs must be addressed, which is accomplished by minimizing communication among individual population members though asynchronous operations. The effect of asynchronous PSO on robustness and efficiency is assessed as a function of problem and population size. Experiments were performed with different population sizes on NVIDIA GPUs and on single-core CPUs. Results for successful trials exhibit marked speedup increases with the population size, indicating that more particles may be used to improve algorithm robustness while maintaining nearly constant time. This work also suggests that asynchronous operations on the GPU may be viable in stochastic population-based algorithms to increase efficiency without sacrificing the quality of the solutions.

1. Introduction

Particle Swarm Optimization (PSO) is a relatively new stochastic global optimization algorithm that is used in a variety of problems, including model calibration and parameter estimation [1–6]. Similar to genetic algorithms and evolutionary strategies, PSO is an iterative, stochastic population-based technique. However, in contrast to genetic algorithms, which model the competitive characteristics of biological evolution, PSO simulates cooperative and social behaviour, such as fish schooling, birds flocking, or insects swarming. A diffuse population of P individuals, now termed as particles, explores the search space, gradually forming smaller swarms in optimal regions. PSO has been successful in many applications, including difficult simulation-based optimization problems [7]. Investigation into the theoretical and convergence properties of PSO is an active, on-going research pursuit [8–10].

PSO offers many advantages. It is relatively easy to conceptualize and to implement, requires few parameters to tune, and, in many cases, demonstrates robust convergence to high-quality solutions. Numerous adaptations and enhancements have been made to PSO, many of which were designed to solve specific optimization problems, or to address shortcomings to which PSO is susceptible, such as

[1] To whom any correspondence should be addressed.

slow convergence and possible entrapment in local optima [8–10]. A particularly useful benefit is that PSO exhibits a high degree of data-level (fine-grained) and task-level (coarse-grained) parallelism. It is therefore conducive to implementation on high-performance computing hardware, and many investigators have explored the potential of parallel PSO on both distributed-memory clusters and on shared-memory supercomputers [11–13].

Within the last ten years, novel parallel architectures have been gaining prominence in high-performance computing. Because processor performance cannot continue to improve indefinitely (due to the inability to reduce the operating voltage of chips as sharply as in the past with increasing numbers of transistors), performance gains are increasingly spearheaded by multicore, graphics processing units (GPUs), and hybrid technologies. GPUs are highly parallel processors that were initially designed to render graphics through a fixed function graphics pipeline. The power of general purpose GPU computation became evident in the mid 2000s, and, consequently, frameworks for GPU programming have been developed, each with unique benefits. Popular GPU environments include CUDA (Compute Unified Device Architecture) by NVIDIA [14], BROOK+ by AMD, and OpenCL by the Khronos group. OpenCL is an open standard for parallel programming of heterogeneous systems that can be used to program the CPU, GPU, Cell architecture, clusters, and even mobile devices, and which supports both data and task-based parallel programming models [15].

In this paper, an asynchronous GPU-based approach to parameter estimation using PSO is presented. Although individual particles simultaneously explore many different regions of the search space (data parallel/fine-grained), because of the complexity of the cost function, and because the cost function often requires observational data, each particle performs significant computation (task parallel/coarse-grained). Consequently, the approach presented here can be termed "medium-grained", where GPU-specific memory bandwidth issues in CPU-GPU communication are addressed. The approach taken in this paper is similar to that taken by other investigators working in parallel PSO on GPUs [16–17]. However, these previous investigations mostly report results on standard global optimization test functions (although some have quite high dimensionality), whereas the current study considers relatively realistic, compute- and data-intensive problems.

2. The Particle Swarm Algorithm

At each PSO iteration, a d-dimensional particle $\mathbf{p}_i$, $i = 1, \ldots, P$, representing a point in the search space, evaluates the cost function $f(\mathbf{p}_i)$. The particle then probes the search space by modifying its position according to a velocity vector, which is a function of the best position found by that particle – its personal best ($\mathbf{p}_i^{best}$) – and of the best position – the global best ($\mathbf{g}^{best}$) – found so far among all particles. In this way, particles tend to congregate (swarm) around good positions. As PSO is stochastic, particles may move on a trajectory in which a better response is encountered. The other particles subsequently adjust their movements to swarm around the better position. A particle's movement is influenced not only by the best global position, but also by its personal best position and by stochastic perturbations. Of the few parameters that PSO does require, the inertial weight w indicates the relative influence of the velocity in the previous iteration and determines the degree to which the particle should move in the same direction as in the last iteration. It can be adjusted adaptively depending upon the progress of the search. This parameter is very important because it regulates the balance between global expansion (large w values) and local refinement [6, 18]. Two other parameters, C_1 (self) and C_2 (swarm) are not critical for the operation of PSO, but tuning these parameters may result in better convergence and avoidance of local optima.

With these parameters, the basic velocity update equation for particle i is given as:

$$\mathbf{v}_i(t+1) = w(t)\mathbf{v}_i(t) + C_1\varphi_1\left(\mathbf{p}_i^{best}(t) - \mathbf{x}_i(t)\right) + C_2\varphi_2\left(\mathbf{g}^{best}(t) - \mathbf{x}_i(t)\right) \quad (1)$$

where $\mathbf{v}_i(t)$ is the velocity of a particle i at time t, $\mathbf{v}_i(t+1)$ is the velocity in the next iteration $t+1$, $\mathbf{p}_i^{best}$ denotes the best position of $\mathbf{p}_i$ thus far in the search, $\mathbf{x}_i(t)$ denotes the particle's current position at time t, and random numbers φ_1, φ_2 ~ $U(0, 1)$ add stochasticity to the update and diversity to the swarm population. After the new velocity is computed, the position of the i-th particle is updated as:

$$\mathbf{x}_i(t+1) = \mathbf{x}_i(t) + \mathbf{v}_i(t+1). \quad (2)$$

3. Methods

3.1. Parallel implementation of PSO

Investigators have previously reported success in exploiting the inherent parallelism of PSO on GPU architectures [16–17]. In this paper, a similar "medium-grained" parallel approach was taken wherein each particle is represented by an individual thread running on the GPU. Separate implementations were developed for single-thread CPU and parallel GPU execution. The CPU and GPU codes are nearly identical, but each particle is given its own individual thread in the GPU implementation.

3.2. Asynchronicity in the GPU implementation

To reduce communication time between GPU cores, asynchronicity was introduced into PSO. Asynchronous parallel PSO has been explored by other investigators [12–13]. In this method, the GPU cores run entirely independently for a specified number of iterations, after which they are re-synchronized. As reported below, it was observed that asynchronicity does not have an adverse effect on the quality of the solutions.

3.3. Adaptive inertial constant

A standard PSO modification allows the inertial parameter of the velocity function w to be modified based upon the number of iterations. In the current work, w was adjusted based on the population's performance. If stagnation is detected (50 iterations have been expended with no improvement in the global best value), the inertial parameter is either increased (to encourage more global exploration) or decreased (to refine existing good solutions). This adjustment requires additional computation time, but has been shown to avoid entrapment in local minima [6, 18].

3.4. CUDA implementation

CUDA is a framework for GPU programming on NVIDIA graphics cards that supports C-like language constructs. It contains functions to simplify the use of the GPU for general purpose processing. At the lowest level, parallelization is achieved via threads, which are analogous to threads in serial computing. GPU programs, called kernels, run in these threads. Blocks, at the next higher level, are simply large arrays of threads. At the highest level, grids are arrays of blocks. CUDA employs symmetric processing: thread tasks are determined according to the block and grid in which the threads reside, as well as their position in these blocks and grids. CUDA also simplifies the transfer of information between RAM and the GPU-specific RAM.

The basic schema for CUDA parallelization consists of three general steps. First, the data to be processed are transferred from the CPU to the GPU RAM. Next, the kernel is run to carry out the processing using the thread number to determine which segment of the data is to be processed by that thread. Finally, the processed data are copied back to the CPU for use and further processing. Additional details of the GPU implementation of PSO are found in [17].

3.5. Cost functions

Three cost functions, representing realistic problems, were explored. Ground truth values for these functions were known.

3.5.1. Toy protein folding. In the toy protein folding problem, intra-molecular potential energy Φ is to be minimized [19]. Each monomer (amino acid) is connected to the next monomer in a protein chain through a bond angle $\theta \in (-\pi/2, \pi/2]$. For a protein of n amino acids, there are $d = n - 2$ such angles. The "toy" model is a simplification of the real protein folding problem in that all monomers are constrained to be co-planar, the bond distances are equal, and only two monomers, designated as A and B, are used. Monomer pairs are restricted to AA, AB/BA, and BB. PSO for this model has been studied by other investigators [20]; however, to the authors' knowledge, the work presented here is the first GPU implementation. The cost function is given as a function of $\boldsymbol{\theta}$:

$$\Phi(\boldsymbol{\theta}) = \sum_{i=2}^{n-1} V_1(\theta_i) + \sum_{i=1}^{n-2} \sum_{j=i+2}^{2} V_2(r_{ij}, \xi_i, \xi_j), \text{ where} \quad (3)$$

$$r_{ij} = \left\{ \left[1 + \sum_{k=i+1}^{j-1} \cos\left(\sum_{l=i+1}^{k} \theta_l \right) \right]^2 + \left[\sum_{k=i+1}^{j-1} \sin\left(\sum_{l=i+1}^{k} \theta_l \right) \right]^2 \right\}^{\frac{1}{2}}, \quad V_1(\theta_i) = \frac{1}{4}(1 - \cos\theta_i), \quad V_2(r_{ij}, \xi_i, \xi_j) = 4\left[r_{ij}^{-12} - C(\xi_i, \xi_j) r_{ij}^{-6} \right]$$

$$\text{and} \quad C(\xi_i, \xi_j) = \left\{ +1, +\frac{1}{2}, -\frac{1}{2} \right\} \text{ for AA, BB, AB pairs.}$$

3.5.2. Logistics function. A simplified, yet realistic two-dimensional logistics problem, used as a benchmark for regression algorithms, was optimized [21]. Calculation requires observational data that must be cached on the GPU to avoid costly global memory accesses. The cost function is not an "analytic" function, as often appears in the global optimization literature (e.g. Rosenbrock, Ackley, Rastigrin functions, etc. [17]), but rather it is a maximum likelihood estimator wherein a regression is found on an individual's age and the incidence of coronary heart disease (CHD). The log likelihood $L(\boldsymbol{\beta})$ is given as a function of $n = 100$ observations with parameters $\boldsymbol{\beta} = [\beta_0\ \beta_1]$ [21]:

$$L(\boldsymbol{\beta}) = \sum_{i=1}^{n} y_i(\beta_0 + \beta_1 x_i) - \sum_{i=1}^{n} \log[1 + \exp(\beta_0 + \beta_1 x_i)], \quad \beta_0 \in (-10, 10), \quad \beta_1 \in (-2, 2), \quad (4)$$

$$x_i = \text{age of patient } i \text{ (years)}, \quad y_i = \begin{cases} 1 & \text{if coronary heart disease in } i\text{th patient} \\ 0 & \text{no CHD} \end{cases}$$

To demonstrate the difficulties involved in optimizing these functions, even in low dimensions, cost-function landscapes of Equations 3 and 4 are shown in figures 1 and 2, respectively. Colour-coded values of the cost function (blue = low, red = high) with varying parameters are shown. In figure 1, The 2D ABAB protein folding cost function values (Eq. 3) are shown for values of the two angles θ_1 and θ_2 between ABA and BAB, respectively, while figure 2 shows all function values for all allowable β_0 and β_1 (Eq. 4). Cost function values are log-scaled to enhance visualization. Figure 1 shows that the toy protein model, $\Phi(\boldsymbol{\theta})$ (Eq. 3), is characterized by long, winding "ridges" of local minima into which a search may become entrapped. The log-likelihood logistics function, $L(\boldsymbol{\beta})$, exhibits a long, narrow valley of local optima that must be overcome.

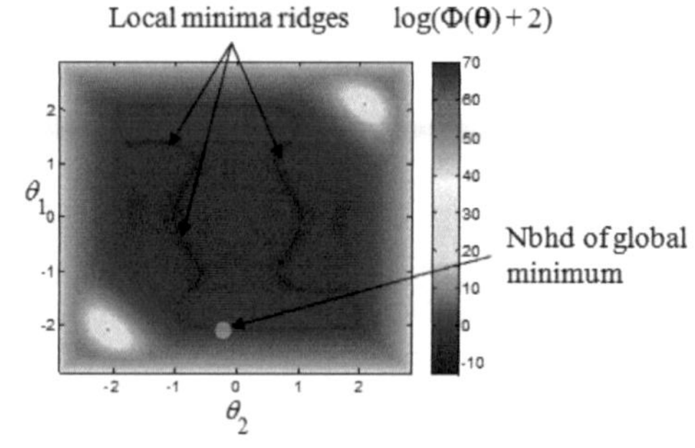

Figure 1. Log-scaled cost function for the 4-monomer ABAB toy protein folding problem, exhibiting "ridges" of local minima.

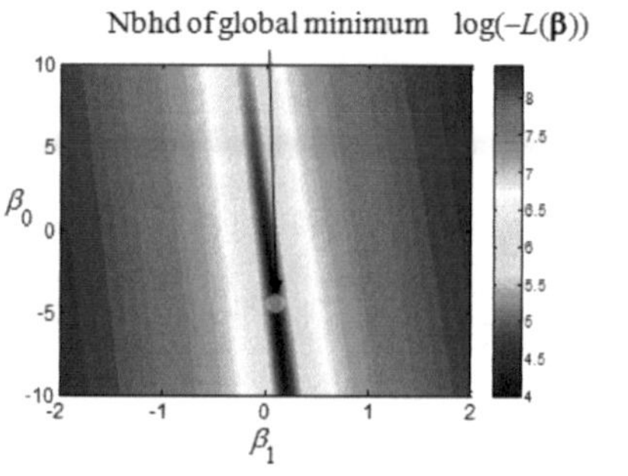

Figure 2. Log-scaled log likelihood cost function for the logistics problem, showing a narrow valley of local minima.

3.5.3. Disequilibrium problem. The disequilibrium problem in econometrics concerns determining the supply and demand components of a time series of transacted quantities (Q). This problem has been studied in the context of global optimization methods [6, 22]. For time period t, the quantity transacted (Q_t), supply (S_t), and demand (D_t) are related by:

$$D_t = \alpha_0 + \alpha_1 X_{1t} + \alpha_2 P_{t-1} + u_t,\ \ S_t = \beta_0 + \beta_1 X_{2t} + \beta_2 P_{t-1} + v_t,\ \ Q_t = \min(D_t, S_t). \tag{5}$$

P_t is the price in period t, X_{1t} and X_{2t} are explanatory variables for period t, and $u_t \sim N(0,\sigma_1)$, and $v_t \sim N(0, \sigma_2)$. The goal is to determine $\boldsymbol{\theta} = [\alpha_0, \alpha_1, \alpha_2, \beta_0, \beta_0, \beta_0, \sigma_1\ \sigma_2]$ by maximizing the log likelihood:

$$L(\boldsymbol{\theta}) = \sum_{t=1}^{T} \log\left(f_{1t}F_{2t} + f_{2t}F_{1t}\right), \text{ where} \tag{6}$$

$$f_{1t} = \frac{1}{\sqrt{2\pi}\sigma_1}\exp\left(-h_{1t}^2/2\right),\ f_{2t} = \frac{1}{\sqrt{2\pi}\sigma_2}\exp\left(-h_{2t}^2/2\right),$$

$$F_{i,t} = \int_{h_{i,t}}^{\infty} \frac{1}{\sqrt{2\pi}}\exp\left(-u^2/2\right)du = \frac{1}{2}\left(1 - \operatorname{erf}\left(\frac{h_{i,t}}{\sqrt{2}}\right)\right),\ \ i = 1,2 \tag{7}$$

$$h_{1t} = \frac{Q_t - \mathbf{X}'_{1t}\boldsymbol{\beta}_1}{\sigma_1},\ h_{2t} = \frac{Q_t - \mathbf{X}'_{2t}\boldsymbol{\beta}_2}{\sigma_2}.$$

3.6. Experiments

Forty (40) experiments were run on each cost function: (1) Three 3D toy protein folding models with monomers AAAAA, AABBB, and BBBBB [19]; (2) The logistics function [21]; and (3) The disequilibrium problem [22]. Ground truth parameters and known global minima are found in [19], [21], and [22] for problems (1) – (3), respectively. All GPU experiments were run on clusters with NVIDIA Tesla S1070 GPUs (1.44 GHz clock speed) with 240 microprocessors supporting double precision computation, and two gigabytes of global memory. CPU results were obtained on a quad-core Intel E5430 Xeon processor with 8GB of FBD PC 5300 RAM. GPU code was developed with the NVIDIA CUDA framework. Code optimizations were similar in both implementations. For asynchronicity, each particle (population member) explored the search space independently and updated the rest of the flock every 50th iteration. Populations of 500, 1000, 2000, and 4000 particles were each budgeted 10000 iterations. The adaptive inertial constant *w* was applied to reduce entrapment in local minima. Set to 1.0 initially, *w* was incremented by 0.01 if there was no improvement in the cost function after 50 iterations, to a maximum of 1.03, to facilitate global exploration. If stagnation still occurred, *w* was reduced by 0.01 to a minimum of 0.97 to refine good areas already identified [6].

4. Results

All protein folding optimizations resulted in correct results on both the GPU and CPU. The logistics function experiments failed with 500 particles, and therefore results for larger populations are presented (97.5% correct for CPU trials, 100% for GPU). The 8D disequilibrium function was the most difficult to optimize, because of the complexity of the cost function landscape and because of the inherent noisiness of the input data [6] (77.5% correct for CPU trials, 80.0% for GPU).

The mean GPU and CPU computation times are shown in Table 1. Using 500 particles as a baseline (1000 for the logistics function), it is seen that computation time for all cost functions increases linearly with the number of particles on the single threaded CPU, as expected – as the number of particles doubles, the computation time also approximately doubles. The computation times on the GPU, as well as being lower, also remain constant with the number of particles, as additional computing resources are not required. More resources on the CPU are required for larger populations, whereas on GPU hardware, existing processing units are better utilized.

Because the robustness of PSO generally increases with the number of particles, speedup was computed based on the population size. That is, for P = 500, 1000, 2000, and 4000, $Speedup(P)$ = CPU time (P) / GPU time (P). The results are shown in figure 3. The toy protein model was a simple calculation of an explicit formula (Eq. 3), and therefore, was very efficient on CPUs, lowering the CPU/GPU speedup. The logistic and disequilibrium functions require more computation and also make use of observed data. The 2D logistic function had the lowest computation time overall, but also the highest CPU/GPU speedup. The 8D disequilibrium problem required the most CPU time. However, if f_{1t}, f_{2t}, h_{1t}, or h_{2t} become unreasonably large (because of division by small σ_1 or σ_2), $L(\theta)$ (Eq. 6) is not considered valid, and the computation does not proceed. In this way, "bad" values are not used in the swarm, thereby increasing efficiency and the CPU/GPU speedup.

Table 1. Mean (± std. dev.) optimization time (seconds) for cost functions by number of particles. GPU results are shown in boldfaced type.

	Particles			
Cost Function	500	1000	2000	4000
Toy Protein AAAAA CPU	125.16 ± 3.06	250.00 ± 5.11	492.50 ± 14.08	990.07 ± 18.75
Toy Protein AAAAA GPU	**9.37 ± 0.13**	**9.27 ± 0.01**	**8.99 ± 0.01**	**9.01 ± 0.01**
Toy Protein BBBBB CPU	121.34 ± 2.20	248.51 ± 3.47	487.33 ± 11.61	990.89 ± 24.92
Toy Protein BBBBB GPU	**9.36 ± 0.01**	**9.25 ± 0.02**	**9.01 ± 0.01**	**8.99 ± 0.01**
Toy Protein AABBB CPU	111.95 ± 0.81	223.42 ± 0.13	446.68 ± 0.20	892.87 ± 1.12
Toy Protein AABBB GPU	**9.32 ± 0.02**	**9.24 ± 0.01**	**8.94 ± 0.00**	**9.12 ± 0.01**
Logistics (log likelihood) CPU	-	99.29 ± 0.06	198.66 ± 0.41	397.02 ± 0.03
Logistics (log likelihood) GPU	**-**	**1.44 ± 0.00**	**1.36 ± 0.00**	**1.33 ± 0.00**
Diseq. (log likelihood) CPU	133.27 ± 6.15	291.11 ± 91.41	534.45 ± 42.37	1137.70 ± 236.83
Diseq. (log likelihood) GPU	**5.30 ± 0.00**	**5.24 ± 0.01**	**5.19 ± 0.01**	**5.29 ± 0.02**

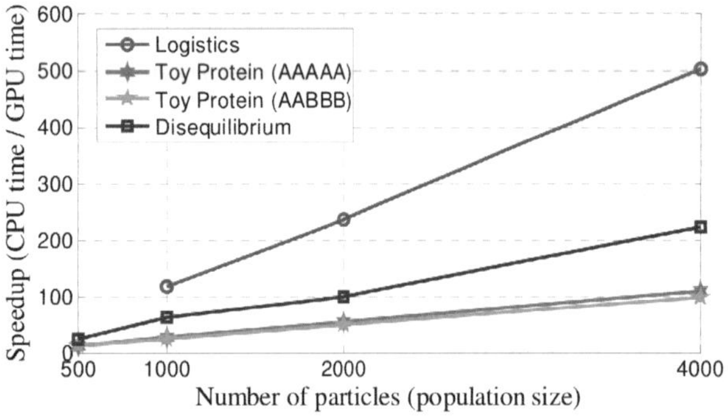

Figure 3. Mean speedup for cost functions (GPU vs. CPU). The speedup for Toy Protein BBBBB is nearly identical to that of AAAAA, and therefore the graph for the former is not shown.

5. Discussion

GPU computing takes advantage of data parallelism, and the current work is within the fine-grained paradigm by assigning one particle per thread. However, the logistics function and the disequilibrium problem require external experimental data – each patient's age and CHD incidence for the former, and quantity transacted and explanatory variables for the latter. Each thread in both cases needed to access values from a large table of memory (originally stored entirely in global memory), which consequently increased the total execution time of the kernel. A more efficient use of the cache and of local memory reduced the slowdown, but longer relative execution times were still observed because memory overhead was added, reducing data parallelism. However, the massive parallelism of the GPU allowed more kernels (particles) to explore the search space simultaneously.

Previously, limiting the number of particles was seen as necessary to avoid slow convergence in serial implementations. With distributed-memory parallelism (e.g. MPI implementations), a reasonable number of particles was required to reduce communications overhead, and with SMP parallelism (e.g. with OpenMP), the number of available threads was an important consideration. With GPUs, however, increasing the number of particles (population size) is an easy way to make best use of this hardware. As seen from Table 1, the running times for each problem remained almost constant (zero slope) with an increasing population size, while the serial implementation, as expected, exhibited a linear increase in computation time. For instance, in the current work, 256 threads (particles) per block can be executed concurrently. There are 30 available processors on the GPU cluster, for a total of 7680 potentially concurrent threads. Therefore, a population of 4000 particles makes better use of resources than one with 1000 particles. Although speedup will decrease as the number of particles exceeds available concurrent threads, even smaller dual- and quad-core systems can take advantage of available GPU capabilities to execute a large number of concurrent threads.

6. Conclusion

Global optimization is one of the important target areas for general purpose GPU computing. Inherently parallel algorithms, including population-based stochastic methods such as PSO, are prime candidates for GPGPU implementation. The results presented in this paper demonstrate that parameter estimation problems requiring external data (contrasted with explicitly defined functions) are also possible on this new architecture without over-burdening the CPU-to-GPU memory bottleneck. Excellent scaling is observed with the population size, as more particles imply better use of available GPU compute resources. Furthermore, asynchronous execution has a minimal effect on the quality of the solutions. For larger parameter estimation problems, managing data and data transfer remain difficult, as larger observed and explanatory data are required. Employing hybrid parallel architectures, such as multicore/many-core CPUs with GPU acceleration, will likely prove valuable, and may assist in overcoming the limits of GPU-only computing.

Although these results are encouraging, it should be noted that the cost functions addressed here were difficult, but relatively low-dimensional. PSO, while easy to code and requiring few tunable parameters, is less robust vis-à-vis other global approaches (such as genetic algorithms) for higher-dimensional problems [23]. Additionally, for realistic model calibration and simulation-based global optimization problems, other stochastic and deterministic approaches may be preferred [22–23]. Consequently, future work will include GPU and hybrid implementations to apply global optimization to solve more data intensive problems, and for realistic model calibration in practical applications.

Acknowledgments

The authors thank Dr. Renata Wachowiak-Smolíková, Jessica Tanner, Bryan Sarlo, and the anonymous reviewers for their helpful comments and criticisms. This work was supported by, and experiments run on the facilities of the Shared Hierarchical Academic Research Computing Network (SHARCNET: www.sharcnet.ca) and Compute/Calcul Canada. Funding was provided by an NSERC Discovery Grant.

References

[1] Poli R, Kennedy J and Blackwell T 2007 *Swarm Intelligence Journal* **1** 33–57
[2] Blum C and Merkle D 2008 *Swarm Intelligence: Introduction and Applications* (New York: Springer-Verlag)
[3] Clerc M 2006 *Particle Swarm Optimization* (California: ISTE Publishing Company)
[4] Engelbrecht A P 2006 *Fundamentals of Computational Swarm Intelligence* (New Jersey: John Wiley & Sons)
[5] Schutte J F and Groenwold A A 2005 *J Global. Optim.* **31** 93–108
[6] Wachowiak M P, Wachowiak-Smolíková R and Smolík D 2010 *Ekonomická revue – Central European Review of Economic Issues* **13** 193-9
[7] Zhang Y, Gallipoli D and Augarde C E 2009 *Computers and Geotechnics* **36** 604–15
[8] Poli R, Kennedy J, Blackwell T and Freitas A (Editors) 2008 *Special Issue. Particle Swarm Optimisation. The Second Decade. Journal of Artificial Evolution and Applications*
[9] Banks A, Vincent J and Anyakoha C 2007 *Natural Computing* **6(4)** 467–84
[10] Banks A, Vincent J and Anyakoha C 2008 *Natural Computing* **7(1)** 109–24
[11] Schutte J F, Reinbolt J A, Fregly B J, Haftka R T and George A D 2004 *Int. J. Numer. Meth. Eng.* **61(13)** 2296–315
[12] Koh B I, George A D, Haftka R T and Fregly B J 2006 *Int. J. Numer. Meth. Eng.* **67** 578–95
[13] Venter G and Sobieszczanki-Sobieksi J 2006 *Journal of Aerospace Computing Information and Communication* **3** 123–37
[14] Nickolls J, Buck I, Garland M and Skadron K 2008 *ACM Queue* **6(2)** 40–53
[15] Stone J E, Gohara D, Shi G 2010 *Comput. Sci. Eng.* **12(3)** 66–73
[16] Veronese L P and Krohling R A 2009 *Proceedings of the Eleventh Conference on Congress on Evolutionary Computation* (CEC'09) 3264–70
[17] Hung Y and Wang W 2012 *Optimization Methods and Software* **27** 33-51
[18] Chatterjee A and Siarry P 2006 *Computers and Operations Research* **33** 859–71
[19] Stillinger F H, Head-Gordon T and Hirshfeld C L 1993 *Phys. Rev.* E **48(2)** 1469-77
[20] Liu J, Wang L, He L, and Shi F 2005 *Proc. ICNC 2005, LNCS 3612* 636–45
[21] Hosmer D W 2000 *Applied Logistic Regression* (New Jersey: John Wiley & Sons)
[22] Jerrell M E, Campione W A 2001 *J. Global. Optim.* **20** 273–95
[23] Erdinc O and Uzunoglu M 2012 *Renew. Sust. Energ. Rev.* **16** 1412–25

Monte Carlo simulation on pre-clinical irradiation: A heterogeneous phantom study on monoenergetic kilovoltage photon beams

James C L Chow

Department of Radiation Oncology, University of Toronto and Radiation Medicine Program, Princess Margaret Hospital, University Health Network, Toronto, Ontario M5G 2M9, Canada; Department of Physics, Ryerson University, Toronto, Ontario, Canada.

james.chow@rmp.uhn.on.ca

Abstract. This study investigated radiation dose variations in pre-clinical irradiation due to the photon beam energy and presence of tissue heterogeneity. Based on the same mouse computed tomography image dataset, three phantoms namely, heterogeneous, homogeneous and bone homogeneous were used. These phantoms were generated by overriding the relative electron density of no voxel (heterogeneous), all voxel (homogeneous) and the bone voxel (bone homogeneous) to one. 360° photon arcs with beam energies of 50 - 1250 keV were used in mouse irradiations. Doses in the above phantoms were calculated using the EGSnrc-based DOSXYZnrc code through the DOSCTP. Monte Carlo simulations were carried out in parallel using multiple nodes in a high-performance computing cluster. It was found that the dose conformity increased with the increase of the photon beam energy from the keV to MeV range. For the heterogeneous mouse phantom, increasing the photon beam energy from 50 keV to 1250 keV increased seven times the dose deposited at the isocenter. For the bone dose enhancement, the mean dose was 2.7 times higher when the bone heterogeneity was not neglected using the 50 keV photon beams in the mouse irradiation. Bone dose enhancement affecting the mean dose was found in the photon beams with energy range of 50 – 200 keV and the dose enhancement decreased with an increase of the beam energy. Moreover, the MeV photon beam had a higher dose at the isocenter, and a better dose conformity compared to the keV beam.

1. Introduction

In radiotherapy, the external-beam treatment planning (clinical or pre-clinical) requires a calculation of radiation dose delivered to the tumour, while sparing the critical or normal tissues from harmful effect. Considering the high accuracy, Monte Carlo simulation is preferred to those semi-experimental or classic approaches in dose calculation [1]. However, Monte Carlo simulation is computation intensive, and usually takes days to finish the dose calculation of a treatment plan [2]. In this work, pre-clinical dosimetric study using Monte Carlo simulation was performed by a high-performance computing cluster to enhance the computing speed, so that a pre-clinical heterogeneous phantom study can be carried out efficiently.

Realistic pre-clinical models for the fractionation, normal tissue dose-volume and radiobiology studies are being developed to support the lack of clinical data in radiotherapy. Since the accuracy of the pre-clinical model depends on the similarity of the small-animal irradiation mimicking the radiotherapy process of the patient, many state-of-the-art radiation irradiators [3-5] with different photon beam energies, image-guided accessories [6], and treatment planning systems [7, 8] for the small animal were developed. The association of the above irradiation hardware and software for the advanced small-animal treatment procedure results in a highly conformal dose to the target volume, and a dose sparing the surrounding healthy tissue as much as possible.

Small-animal irradiation and imaging accessories such as the computed tomography (CT) [9-11], magnetic resonance [12-14] and position emission tomography [15-17] imaging are available in the mouse scale of millimeter. In order to further match the size of mouse compared to human, megavoltage (MV) photon beams (4 – 18 MV) generally used in the patient's radiotherapy are replaced by kilovoltage (kV) beams because of the relatively higher beam attenuation. The use of kV photon beams in the mouse irradiation has dosimetric reasons of the beam attenuation and depth dose that fit the size of mouse. However, such kV photon beams in the small-animal irradiation produce extra bone dose, which is not present in the patient's radiotherapy using MV beams. This unwanted bone dose enhancement from the mouse irradiation may lead to an uncertainty in the pre-clinical model. To date, patient dosimetry (e.g. mean and maximum imaging dose) due to heterogeneities such as bone regarding the popular imaging guidance procedures of kV cone-beam CT (kV-CBCT) in image-guided radiotherapy has been studied [18-22]. It is found that because of the increased photoelectric effect for the kV photon beams compared to the MV, a kV-CBCT procedure results in a bone dose enhancement up to a factor of 3 - 4 times than those in the surrounding soft tissue [19, 23]. This increased bone dose would affect the bone marrow which is a radiosensitive organ-at-risk [24-27]. However, there is a lack of studies concerning the bone dose enhancement for the small animal when using kV photon beams in the irradiation.

Although there are studies on dosimetric variations of pre-clinical irradiations using polyenergetic photon beams produced by different mouse irradiators [28, 29], there is no investigation concerning dosimetric variations using monoenergetic photon beams. In this study, a mouse phantom which is heterogeneous based on a mouse CT image dataset was used in Monte Carlo simulation and irradiated with monoenergetic kV photon beams with different energies. To study the effect of the bone dose enhancement due to the kV photon beam irradiation, the relative electron density of every bone voxel in the above phantom was overridden to one in order to generate another bone homogeneous mouse phantom without the bone heterogeneity. Monte Carlo simulations were performed by sampling variables of photon and electron energy depositions. Simulations on irradiations of the 360° photon arcs with energies of 50, 100, 150, 200, 250, 300 and 1250 keV were carried out in the above two phantoms together with a homogeneous mouse phantom with all relative electron densities in voxels overridden to one. The Electron Gamma Shower (EGSnrc)-based code was used to calculate doses in the three mouse phantoms [30, 31]. Monte Carlo simulation of particle transports in heterogeneous media is currently the most accurate dose calculation method for kV photon beams with different atomic numbers [5, 32]. The DOSCTP graphical user interface based on the DOSXYZnrc code was used to carry out the treatment plan and phantom edition [8, 33]. The aim of this study is to investigate dosimetric variations in the presence and absence of tissue heterogeneities, in particular the bone, in mouse irradiations using monoenergetic photon beams with different energies in the kV range. Doses at the interesting points (the isocenter and bone), mean doses, and maximum doses were calculated by Monte Carlo simulations using high-performance computing cluster.

2. Methods

A mouse CT image dataset obtained by a 225 kVp x-ray tube (GE Siefert 225, focal spot size = 0.4 mm) and flat-panel detector (Perkin Elmer, IRD1640) was converted to an EGSnrc-based Monte Carlo phantom using a modified CTCREATE routine with the calibrated CT number and density conversion ramp specified for the mouse [7, 8]. The voxel size of the phantom was set to $0.1 \times 0.1 \times 0.18$ cm^3 for the x, y and z-axis, respectively. The z-axis of the phantom is parallel to the superior-inferior direction of the mouse. Figure 1(a) shows the x-y plane of the Monte Carlo heterogeneous mouse phantom with the isocenter, rib and spine. Each division in the ruler scale represents 1 cm in Fig. 1. Based on the heterogeneous mouse phantom in Fig. 1(a), a homogenous mouse phantom (Fig. 1(b)) was then generated by overriding the relative electron density in every voxel to one using the phantom editing feature in the DOSCTP code. In addition, a bone homogeneous mouse phantom as shown in Fig. 1(c) was generated by overriding the relative electron density of every bone voxel in the inhomogeneous mouse phantom to one. Through dosimetric comparisons among the heterogeneous, homogeneous and bone homogeneous mouse phantoms irradiated by different energies of photon beams, the effect of the presence of the heterogeneity such as bone in the dose distribution, which varies with the beam energy, can be investigated. To study the strength difference of the photoelectric effect in the bone between the kV and MV photon beams, a beam of 1250 keV was used to irradiate the three phantoms for comparison.

In radiation dose calculation using Monte Carlo simulation, paths of the individual particles (photons and electrons) are governed by random numbers distributed according to the known probabilities of the various scattering and energy deposition events. The energy deposition events are collected in cubical bins, and the energy deposited per unit mass is taken as a measure of dose that would be deposited under actual irradiation conditions. Since each particle is by definition statistically independent as well as governed by the same process, Monte Carlo simulations can be classified as "embarrassingly parallel" problem when implemented. As the accuracy of Monte Carlo simulation depends on the sampling size (histories), and the typical number of histories used for an accurate dose calculation (statistical uncertainty = ± 1%) is equal to couple hundred million, Monte Carlo simulation is recognized as a time-consuming but accurate method in dose calculation. Therefore, high-performance computing is needed to improve the simulation speed. There are some studies concerning the speed enhancement in Monte Carlo simulation. Hissoiny et al [34] developed a new Monte Carlo dose calculation platform called "GPUMCD" using graphical processing unit (GPU). The GPUMCD can perform Monte Carlo radiation dose calculation for energies in the range of 0.01 – 20 MeV using the class II condensed history method. GPUMCD runs on a NVIDIA GTX480 graphical processor, and dosimetric results from GPUMCD were validated by the EGSnrc code that over 98% of voxels passed the gamma criteria of 2%/2mm. The computing speed of GPUMCD is more than 900 times faster than the EGSnrc running on a typical computing processor. Chow et al [35] investigated the probability to implement Monte Carlo simulation using the cell processor in the Playstation3 console. They found that the EGSnrc code could be parallelizable on the cell processor, provided that an architectural change of the EGSnrc was made so that the code was initiated on the powerPC processing element, and execution of the simulation loop within the program be split between the PowerPC processing and synergistic processing elements in the system. However, the performance of cell processor is worse than the GPU due to the limitation of the PowerPC processing element. In this study, the EGSnrc code version 4-r2-2-5 developed by the National Research Council Canada was used [30, 31]. The code was run in parallel using multiple nodes on the General Purpose Cluster at SciNet from the University of Toronto. The computing time was about 10 – 20 minutes per calculation, which is faster than a single PC CPU with computing time equal to 2 – 3 hours per calculation. The DOSXYZnrc associated with the EGSnrc was used to calculate the dose [33], and the DOSCTP running on the MATLAB platform was used to create the Monte Carlo mouse phantoms and irradiation plans [7].

Monoenergetic photon beams with energies of 50, 100, 150, 200, 250, 300 and 1250 keV were used in Monte Carlo simulations. The field size of the photon beam was 2.5 × 2.5 cm^2. The distance between the field and isocenter was equal to 5 cm as shown in Fig. 1(a). A 360° photon arc around the z-axis was used to irradiate the heterogeneous, homogeneous and bone homogeneous mouse phantom. The input file of the irradiation plan for Monte Carlo simulation was created by the DOSCTP, and the simulation was then automatically run on the DOSXYZnrc linked to the DOSCTP. Sampling histories of five hundred million of particle (photon and electron) tracks were used in each dose calculation. Under this number of history, the relative dose error (i.e. uncertainty as a fraction of the dose in the voxel) was on the order of $\times 10^{-2}$ (1% - 3%) according to the Monte Carlo output file [33]. In dose calculations, the electron transport cut-off energy of 512 keV was used. The photon cut-off energy was set to 1 keV in the phantoms and the maximum fractional electron energy loss per step was set to 25%.

Based on the Monte Carlo results, point doses of interest such as the isocenter, rib and spine were determined from the heterogeneous, homogeneous and bone homogeneous mouse phantoms. Moreover, mean and maximum doses for each of the three phantoms were calculated with different photon beam energies. Although dose points in Fig. 1(a) are not shown in Figs. 1(b) and 1(c), the positions and coordinates of points for the homogeneous and bone homogeneous mouse phantom are the same as the heterogeneous phantom, because all mouse phantoms were generated from the same source of CT image dataset.

3. Results and discussion

Dose distribution of the homogeneous mouse phantom can be seen in Fig. 2 for photon beams with relatively low energy of 50 keV. The dose distribution in Fig. 2 shows the dose deposited by the 360° photon arc is conformed to the isocenter of the phantom, and all doses were normalized to that at the isocenter of the phantom. The diameter of the 100% dose contour in Fig. 2 is about 2.5 cm, which matched the field size of the photon beam. There is a dose gradient from the isocenter to the external contour of the mouse. This dose gradient was found to increase with the increase of photon beam energies from 50 keV to 1250 keV. It shows that higher photon beam energy can increase the dose conformity to the target at the isocenter.

Point doses of the isocenter, rib and spine for the heterogeneous, homogeneous and bone homogeneous mouse phantom irradiated by photon beams with energies in the range of 50 – 1250 keV are shown in Table 1. In the Table, the dose at the isocenter in the homogeneous mouse phantom is only slightly higher than that in the bone homogeneous phantom of about 1.5% for all photon beam energies. Comparing point doses of the rib and spine between the heterogeneous and bone homogeneous mouse phantom in Table 1, bone dose enhancements of 3.4 and 2.9 times are found in the rib and spine when using the 50 keV photon beams. These dose enhancements are due to the photoelectric effect occurring in the keV photon beam energy range in the bone, and are not seen in the 1.25 MeV photon beams. For this MeV photon beams, point doses of the rib and spine in the heterogeneous mouse phantom are 6.5% and 4.5% smaller than those in the bone homogeneous phantom. Moreover, the effect of the bone dose enhancement disappears when the photon beam energy is increased from 50 to 250 keV.

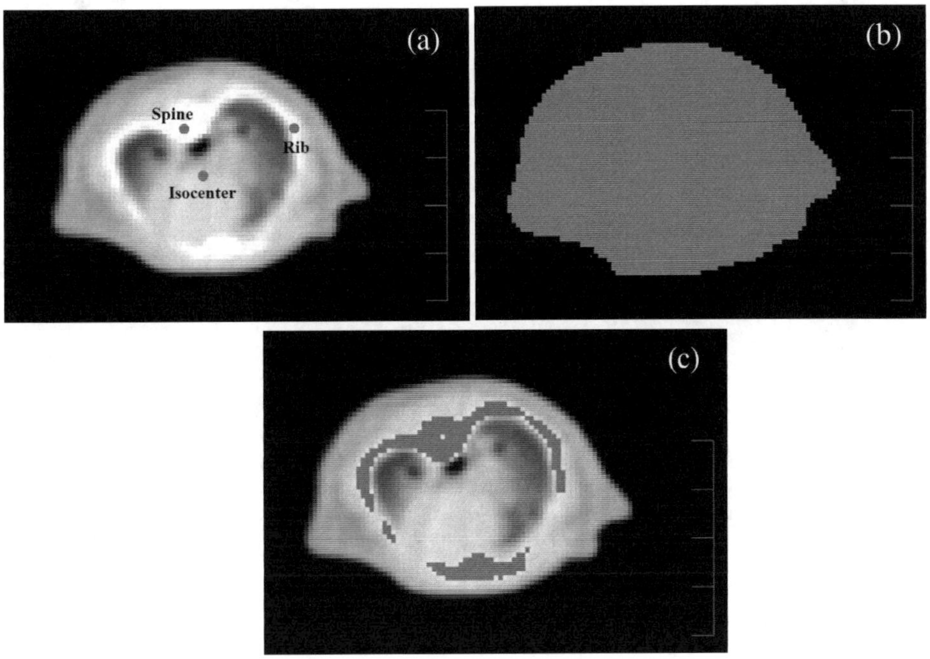

Figure 1. Computed Tomography images of the mouse with the relative electron density of (a) voxel unchanged (heterogeneous mouse), (b) all voxel overridden to 1 (homogeneous mouse), and (c) only the bone voxel overridden to 1 (bone homogeneous mouse). Dose points for the isocenter, rib and spine are shown in 1(a). The ruler scales show 1 cm division.

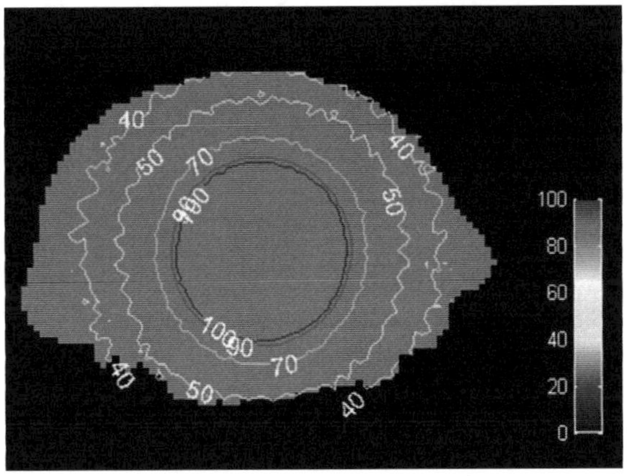

Figure 2. Dose distribution displayed as 2D contours in the homogeneous mouse phantom. Doses were calculated using Monte Carlo simulation with photon beam energy equal to 50 keV.

Table 1. Point doses of the isocenter, rib and spine, as shown in Fig. 1(a), in the heterogeneous, homogeneous and bone homogeneous mouse phantom for photon beam energies of 20, 100, 150, 200, 250, 300 and 1250 keV.

	Isocenter			Rib			Spine		
Photon beam energy (keV)	Hetero mouse (Gy ×10 $^{-13}$)	Homo mouse (Gy ×10 $^{-13}$)	Bone homo mouse (Gy ×10 $^{-13}$)	Hetero mouse (Gy ×10 $^{-13}$)	Homo mouse (Gy ×10 $^{-13}$)	Bone homo mouse (Gy ×10 $^{-13}$)	Hetero mouse (Gy ×10 $^{-13}$)	Homo mouse (Gy ×10 $^{-13}$)	Bone homo mouse (Gy ×10 $^{-13}$)
50	7.34	2.63	2.43	4.36	1.22	1.28	7.52	2.49	2.56
100	5.83	3.23	3.19	2.72	1.49	1.51	5.72	3.22	3.11
150	6.54	5.44	5.27	2.77	2.25	2.31	6.15	5.19	5.33
200	7.93	7.69	7.61	3.41	3.17	3.28	8.01	7.66	7.57
250	10.21	10.13	10.03	4.31	4.27	4.29	9.98	10.12	10.03
300	12.36	12.46	12.34	5.12	5.34	5.18	11.86	12.68	12.48
1250	50.19	53.09	52.48	20.38	22.13	21.79	49.05	51.76	51.35

The variations of the mean and maximum doses in the mouse phantoms due to the heterogeneities for different photon beam energies are shown in Table 2. Comparing the mean doses in the heterogeneous and bone homogeneous mouse phantom, the mean dose in the heterogeneous mouse phantom is about 2.7 times higher than that in the bone homogeneous phantom using the 50 keV photon beams. This mean dose is increased from 50 to 250 keV and then slightly decreased, when the photon beam energy is higher than 250 keV. For the 300 keV photon beams, the mean dose in the heterogeneous phantom is slightly lower than that in the bone homogeneous phantom. This decrease of mean dose is more obvious for the 1250 keV photon beams in which the mean dose in the bone homogeneous mouse phantom is about 2.3% higher than that in the heterogeneous phantom. It is seen that the bone dose enhancement effect due to the keV photon beam is strongest at 50 keV up to 200 keV. Comparing mean and maximum doses of the heterogeneous and homogeneous mouse phantom as shown in Table 2, dose deviations are more significant in the low-energy photon beam (e.g. 50 keV) than the higher (e.g. 1250 keV). The deviations of mean and maximum dose between the heterogeneous and homogeneous mouse phantom due to the heterogeneities are 60.9% and 69.1% for the 50 keV photon beams, and only -1.25% and 1.07% for the 1250 keV photon beams, respectively. These large dose differences between the low- and high-energy photon beams reflect that the dependence of heterogeneities on the dose distribution is more significant in the mouse irradiation using photon beam with energy in the keV range. For photon beam energy in the range of 50 – 150 keV, the deviations of mean and maximum dose between the homogeneous and bone homogeneous mouse phantom are smaller than those between the heterogeneous and bone homogeneous phantom. This shows that the bone dose enhancement due to the photoelectric effect is dominant in this photon beam energy range.

Table 2. Mean and maximum doses in the heterogeneous, homogeneous and bone homogeneous mouse phantom irradiated by the 360° photon arcs with beam energies of 20, 100, 150, 200, 250, 300 and 1250 keV.

	Heterogeneous Mouse		Homogeneous Mouse		Bone homogeneous Mouse	
Photon beam energy (keV)	Mean dose (×10^{-13} Gy)	Maximum dose (×10^{-13} Gy)	Mean dose (×10^{-13} Gy)	Maximum dose (×10^{-13} Gy)	Mean dose (×10^{-13} Gy)	Maximum dose (×10^{-13} Gy)
50	2.61	8.73	1.02	2.70	0.97	2.67
100	1.48	6.05	1.11	3.37	1.11	3.33
150	1.52	6.58	1.42	5.52	1.42	5.42
200	1.98	8.29	1.92	7.96	1.96	7.89
250	2.52	10.39	2.49	10.55	2.54	10.36
300	3.07	12.78	3.05	13.14	3.12	13.00
1250	6.36	55.32	6.44	54.73	6.51	53.94

4. Conclusions

The dosimetric dependence on the photon beam energy and heterogeneity for the mouse irradiation was studied using Monte Carlo simulation. Monte Carlo mouse phantoms namely, heterogeneous, homogeneous and bone homogeneous based on the same mouse CT image dataset were generated by overriding the relative electron density of no voxel, all voxel and the bone voxel to one, respectively. Photon beam energies of 50, 100, 150,

200, 250, 300 and 1250 keV were used for the 360° photon arc in the mouse irradiations. Monte Carlo results showed that dose conformity increased when the photon beam energy increased from 50 keV to 1250 keV, showing that a greater dose gradient could be achieved in higher photon beam energy. In addition, point doses at the rib and spine were found to be 3.4 and 2.9 times higher when comparing the bone doses in the heterogeneous and bone homogeneous mouse phantoms using the 50 keV photon beam. The mean dose in the mouse irradiation was 2.7 times higher when the bone density was overridden to one using the 50 keV photon beam, and the increase of dose was decreased as the beam energy was increased. The dosimetry data in this study is useful when considering the options of photon beam energy in the small-animal treatment planning, or designing a mouse irradiator for pre-clinical model. Moreover, high-performance computing has been used to enhance the dose calculation speed from hours to minutes in Monte Carlo simulation, resulting in a more efficient preclinical study.

Acknowledgement
The author would like to thank Michael Leung at the University of Toronto who prepared the Monte Carlo mouse phantoms using the DOSCTP.

References
[1] Rogers, D W O. 2006 Fifty years of Monte Carlo simulations for medical physics, Phys. Med. Biol. **51**, R287-R301.
[2] Downes P, Yaikhom G, Giddy J P, Walker D W, Spezi E, Lewis D G. 2009 High-performance computing for Monte Carlo radiotherapy calculations, Phil. Trans. R. Soc. A **367**, 2607-2617.
[3] Stojadinovic S, Low D A, Hope A J, Vicic M, Deasy J O, Cui J, Khullar D, Parikh P J, Malinowski K T, E. Izaguirre W, Mutic S and Grigsby P W. 2007 MicroRT – small animal conformal irradiator, Med. Phys. **34**, 4706-4716.
[4] Stojadinovic S, Low D A, Vicic M, Mutic S, Deasy J O, A. Hope J, Parikh P J and Grigsby P W. 2006 Progress toward a microradiation therapy small animal conformal irradiator, Med. Phys. **33**, 3834-3845.
[5] Bazalova M, Zhou H, Keall P J and Graves E E. 2009 Kilovoltage beam Monte Carlo calculation in submillimeter voxel for small animal radiotherapy, Med. Phys. **36**, 4991-1999.
[6] Jaffray D, Moseley D, Chow J, Kim S, Ansell S, Wilson G and Chiarot C. 2006 An image-guided irradiator for preclinical radiation therapy studies, Med. Phys. **33**, 2241.
[7] Chow J C L and Leung M K K. 2007 Treatment planning for small animal using Monte Carlo simulation, Med. Phys. **34**, 4810-4817.
[8] Chow J C L and Leung M K K. 2008 A graphical user interface for the external beam treatment planning using Monte Carlo simulation, J. Phys. CS. **102**, 1-6.
[9] Johnston S M, Johnson G A and Badea C T. 2008 Geometric calibration for a dual tube/detector micro-CT system, Med. Phys. **35**, 1820-1829.
[10] Paulus M J, Gleason S S, Kennel S J, Hunsicker P R and Johnson D K. 2000 High resolution x-ray computed tomography: An emerging tool for small animal cancer research, Neoplasia, **2**, 62-70.
[11] Cavanaugb D, Johnson E, Price R E, Kurie J, Travis E L and Cody D D. 2004 In vivo respiratory-gated micro-CT imaging in small-animal oncology models, Molecular Imaging, **3**, 55-62.
[12] Natt O, Watanabe T, Boretius S, Radulovic J, Frahm J and Michaelis T. 2002 High-resolution 3D MRI of mouse brain reveals small cerebral structures in vivo, Journal of Neuroscience Methods, **120**, 203-209.
[13] Benveniste H and Blackband S. 2002 MR microscopy and high resolution small animal MRI: applications in neuroscience research, Progress in Neurobiology, **67**, 393-420.
[14] Beuf O, Jaillon F and Jaint-Jalmes H. 2006 Small-animal MRI: signal-to-noise ratio comparison at 7 and 1.5 T with multiple-animal acquisition strategies, Magnetic Resonance Materials in Physics, Biology and Medicine, **19**, 202-208.
[15] Correia J A, Burnham C A, Kaufman D, Fischman A J. 1999 Development of small animal PET imaging device with resolution approaching 1 mm, IEEE Trans. Nucl. Sci. **46**, 631-635.
[16] Myers R. 2001 The biological application of small animal PET imaging, Nuclear Medicine and Biology, **28**, 585-593.
[17] Funk T, Sun M and Hasegawa B H. 2004 Radiation dose estimate in small animal SPECT and PET, Med. Phys. **31**, 2680-2686.
[18] Islam M K, Purdie T G, Norrlinger B D, Alasti H, Moseley D J, Sharpe M B, Siewerdsen J H, and Jaffray D A. 2006 Patient dose from kilovoltage cone beam computed tomography imaging in radiation therapy, Med. Phys. **33**, 1573 – 1582.

Journal of Physics: Conference Series **385** (2012) 012013 doi:10.1088/1742-6596/385/1/012013

[19] Ding G X, Duggan D M, and Coffey C W. 2008 Accurate patient dosimetry of kilovoltage cone-beam CT in radiation therapy, Med. Phys. **35**, 1135 – 1144.

[20] Ding G X, Pawlowski J M, and Coffey C W. 2008 A correction-based dose calculation algorithm for kilovoltage x rays, Med. Phys. **35**, 5312 – 5316.

[21] Chow J C L, Leung M K K, Islam M K, Norrlinger B D, and Jaffray D A. 2008 Evaluation of the effect of patient dose from cone beam computed tomography on prostate IMRT using Monte Carlo simulation, Med. Phys. **35**, 52 – 60.

[22] Chow J C L. 2009 Cone-beam CT dosimetry for the positional variation in isocenter: A Monte Carlo study, Med. Phys. **36**, 3512-3520.

[23] Verhaegen F, Schulze C, Seuntjens J, Gosselin M, Hristov D and Svatos M. 2004 Heterogeneity corrected convolution dose calculation for kV x-rays verus Monte Carlo simulation, Med. Phys. **31**, 1770.

[24] Richardson R B, Nie H L and Chettle D R. 2007 Monte Carlo simulation of trabecular bone remodeling and absorbed dose coefficients for tritium and ^{14}C, Rad. Prot. Dosim. **127**, 158-162.

[25] Till J E and McCulloch E A. 1961 A direct measurement of the radiation sensitivity of normal mouse bone marrow cells, Rad. Res. **14**, 213-222.

[26] Theise N D, Badve S, Saxena R, Henegariu O, Sell S, Crawford J M and Krause D S. 2003 Derivation of hepatocytes from bone marrow cells in mice after radiation-induced myeloablation, Hepatology, **31**, 235-240.

[27] Walters B R B, Ding G X, Karmer R, and Kawrakow I. 2009 Skeletal dosimetry in cone beam computed tomography, Med. Phys. **38**, 2915-2922.

[28] Chow J C L, Leung M K K, Lindsay P E, and Jaffray D A. 2010 Dosimetric variation due to the photon beam energy in the small-animal irradiation: A Monte Carlo study, Med. Phys. **37**, 5322-5329.

[29] Bazalova M, Graves E E. 2011 The importance of tissue segmentation for dose calculations for kilovoltage radiation therapy, Med. Phys. **38**, 3039-3049.

[30] Nelson W R, Hirayama H and Rogers D W O. 1965 The EGS4 code system," SLAC Report 265, Stanford Linear Accelerator Center, Stanford, CA.

[31] Kawrakow I and Rogers D W O. 2000 The EGSnrc code system: Monte Carlo simulation of electron and photon transport," NRCC Report PIPRS-701, National Research Council of Canada, Ottawa.

[32] Ding G X, Duggan D M and Coffey C W. 2007 Characteristics of kilovoltage x-ray beams used for cone-beam computed tomography in radiation therapy, Phys. Med. Biol. **52**, 1595-1615.

[33] Ma C M, Reckwerdt P, Holmes M, Rogers D W O and Geiser B. 1995 DOSXYZ user manual, NRC Report PIRS 509b.

[34] Hissoiny S, Ozell B, Bouchard H, Despres P. 2011 GPUMCD: A new GPU-oriented Monte Carlo dose calculation platform, Med. Phys. **38**, 754-764.

[35] Chow J C L, Lam P, Jaffray D A. 2012 Monte Carlo dose calculation using a cell processor based PlayStation 3 system, J. Phys.: Conf. Ser. **341** 012028.

High Performance Computing Symposium 2012 (HPCS2012) IOP Publishing
Journal of Physics: Conference Series **385** (2012) 012014 doi:10.1088/1742-6596/385/1/012014

Plain Polynomial Arithmetic on GPU

Sardar Anisul Haque, Marc Moreno Maza
University of Western Ontario, Canada
E-mail: shaque4@uwo.ca, moreno@csd.uwo.ca

Abstract. As for serial code on CPUs, parallel code on GPUs for dense polynomial arithmetic relies on a combination of asymptotically fast and plain algorithms. Those are employed for data of large and small size, respectively. Parallelizing both types of algorithms is required in order to achieve peak performances. In this paper, we show that the plain dense polynomial multiplication can be efficiently parallelized on GPUs. Remarkably, it outperforms (highly optimized) FFT-based multiplication up to degree 2^{12} while on CPU the same threshold is usually at 2^6. We also report on a GPU implementation of the Euclidean Algorithm which is both work-efficient and runs in linear time for input polynomials up to degree 2^{18}, thus showing the performance of the GCD algorithm based on systolic arrays.

1. Introduction

Until the advent of multicore architectures, algorithms subject to effective implementation on personal computers were often designed with *algebraic complexity* as the main complexity measure and with sequential running time as the main performance counter [15, 16, 17, 4, 10]. Nevertheless, during the past 40 years, the increasing gap between memory access time and CPU cycle time, in favor of the latter, brought another important and practical efficiency measure: *cache complexity* [14, 8]. In addition, with parallel processing becoming available on every desktop or laptop, the *work* and *span* of an algorithm expressed in the fork-join multithreaded model [9, 4] have become the natural quantities to compute in order to estimate *parallelism.*

These complexity measures (algebraic complexity, cache complexity, parallelism) are defined for computation models that largely simplify reality. On multicore architectures, several phenomena (memory wall, true/false sharing, scheduling costs, etc.) limit the performances of applications which, theoretically, have a lot of opportunities for concurrent execution. One infamous example are *Fast Fourier Transforms (FFTs).* For this type of calculation, not only the memory access pattern, but also the for-loop parallelization overheads, restrict linear speedup to input vectors of very large sizes, say 2^{20}, according to [21, 22]. In contrast, serial FFT code provide high-performance even for input vectors of relatively small sizes, say 2^{10}. This is the case with the standard libraries FFTW [7], NTL [25] and Spiral [24]. As a consequence, higher level algorithms, that heavily rely on FFTs in their serial implementation, require additional supporting routines for small/average size problems, when targeting implementation on multicore architectures. Examples of such higher level algorithms are fast evaluation and fast interpolation based on sub-product tree techniques, see Chapter 10 in [10].

Graphics processing units (GPUs) offer a higher level of concurrent memory access than multicore architectures. Moreover, thread scheduling is done by the hardware, which reduces for-loop parallelization overheads significantly. Despite of these attractive features, and as reported in [19], highly optimized FFT implementation on GPUs are not sufficient to support the parallelization of higher level algorithms, such as dense univariate polynomial arithmetic. To give a figure, for vectors of size 2^{18} and 2^{26}, speedup factors (w.r.t. a C serial implementation) are 9 and 37 respectively on a NVIDIA Geforce GTX 285 running CUDA 2.2, as reported in [19].

Consider now the fundamental application of polynomial arithmetic: *solving systems of non-linear equations*. Many polynomial systems encountered in practice have finitely many solutions. Moreover, those systems that can be solved symbolically by computer algebra software, such as MAPLE or MATHEMATICA, have rarely more than 10,000 solutions, see for instance [6]. For this reason, the degrees of univariate polynomials that arise in practice rarely exceed 10,000.

It follows from the above discussion that implementation a polynomial system solver on multicores or GPUs require efficient parallel polynomial arithmetic in relative low degrees, that is, within degree ranges where FFT-based methods do not apply. The study conducted in [3] show that univariate arithmetic based on parallel versions of the Algorithm of Karatsuba and its variants are not effective either in the desired degree ranges. This leads us to consider quadratic (or plain) algorithms for dense univariate multiplication, division and GCD computation. That is, algorithms which run within $O(d^2)$ coefficient operations for polynomials of degree less than d, meanwhile FFT-based algorithms amount to $O(d\log(d)\log(\log(d)))$ coefficient operations, see the landmark textbook [10] for details.

In the present paper, we show that a GPU implementation (with CUDA) of the *plain* univariate multiplication can outperform for fairly large degrees an optimized GPU implementation (with CUDA also) of an FFT-based univariate multiplication. We also report on a GPU implementation of the *plain* univariate division. Section 3 and 4 are dedicated to these multiplication and division CUDA codes, respectively: both contain implementation details, theoretical analysis and experimental results.

We focus on dense polynomial arithmetic over finite fields, since the so-called *modular methods*, such as that presented in [6], allows us to reduce the solving of polynomial systems with rational number coefficients to the solving of polynomial systems over finite fields. To this end, we have realized a preliminary implementation GPU-supported solver for polynomial systems over finite fields. In [19], we report experimental results showing speedup factors of $7,5$ (w.r.t. a highly optimized serial implementation in C) for bivariate systems with $14,400$ solutions. This limited speedup is due to the fact that one essential operation of this solver is performed on the host (and thus is not parallelized yet): univariate polynomial GCD computation.

For this reason, in a third part of the present paper, we turn our attention to one of the most challenging parallelization problem in polynomial arithmetic: the *Euclidean Algorithm*. Indeed, there is no parallel version of this algorithm which would be both sublinear and work-efficient[1]. The best parallel version of the Euclidean Algorithm which is work-efficient, is that for systolic arrays, a model of computation formalized by H. T. Kung and C. E. Leiserson [18], for which the span is linear [2]. Multiprocessors based on systolic arrays are not so common (as they are quite specialized to certain operations and difficult to build). However, the recent development of hardware acceleration technologies (GPUs, field-programmable gate arrays, etc.) has revitalized researchers interest in systolic algorithms and more generally in optimizing the use of computer resources for low-level algorithms [1, 11, 13].

We report on a GPU implementation of the Euclidean Algorithm which is work-efficient and which runs in linear time for input polynomials up to degree 2^{18}. As mentioned above, such sizes are sufficient for many applications. Moreover, our GPU code outperforms algorithms for polynomial GCD computations that are asymptotically faster (such as the Half-GCD algorithm, see [26]) but available only as serial CPU code, due to the same parallelization challenges as the Euclidean Algorithm.

[1] Here work-efficient refers to a parallel algorithm in the PRAM model for which the maximum number of processors in use times the span is in the same order as the work of the best serial counterpart algorithm.

2. A manycore machine model

We describe the computation model that we use for analyzing the algorithms that we implement on GPUs. We follow the CUDA execution model with a few natural simplifications. First, we assume that the number of streaming multiprocessors is unlimited. Moreover, we assume that the global memory is unlimited and support concurrent reads and concurrent writes. However, the shared memory and the number of registers of each streaming multiprocessor remain finite and both regarded as small. In addition, moving data between streaming multiprocessors and the global memory remains a penalty, similarly to what happens on a GPU.

On this ideal machine, programs are similar to CUDA programs. They interleave serial code consisting of C-like code and kernel calls consisting of SIMD code (single instruction multiple data) organized into thread blocks. However, all thread blocks of a given kernel start simultaneously and run concurrently. In addition, data transfer between global memory and shared memories occur only at the beginning or at the end of the execution of a thread block.

This latter two restrictions make algorithms analysis simpler. Given, an algorithm and an input data of size n, running our *ideal manycore machine*, we are interested in the *work* (that is, the total number of arithmetic operations performed by the streaming multiprocessors) the *span* (that is, the running time of the program ignoring the serial code and host-device data transfer) and the number of kernel calls. For these three measures, the lower the better.

3. Plain Multiplication on the GPU

Consider two univariate polynomials over a finite field

$$a = a_n x^n + \cdots + a_1 x + a_0 \text{ and } b = b_m x^m + \cdots + b_1 x + b_0, \text{ with } n \geq m. \quad (1)$$

We start from the so-called *long multiplication*[2], which computes the product $a \times b$ in the way we all learned integer multiplication in primary school. To parallelize it we proceed as follows.

- Multiplication phase: The set of terms $\{a_i b_j x^{i+j} \mid 0 \leq i \leq n, 0 \leq j \leq m\}$ forms a parallelogram that we decompose into small rectangles and triangles such that each of them can be computed by a thread block. Within a thread block, a thread computes all the terms of a given degree range and adds them up into a vector, where this *vector* is associated to the thread block.
- Addition phase: All the thread block vectors are added together by means of a parallel reduction.

Note that in this process, every coefficient of one polynomial is multiplied with every coefficient of the other polynomial. These products of coefficients are independent of each other and, thus, can be done in parallel essentially in $O(1)$ time (if we have sufficiently many processors and enough space to store mn coefficients). These coefficient products can be added in $log(m)$ parallel steps. So the overall complexity of this parallel algorithm is dominated by that of the addition phase. Though the algorithm that we just described seems not realistic due to the limited computing resources of today's computers, still we can adapt its principle in such a way that we make the best use of streaming multiprocessors of GPUs.

3.1. Implementation

In order to assign each thread block with the products $a_i b_j x^{i+j}$ that this thread block performs, we view $a_i b_j x^{i+j}$ as the point of coordinates (i, j) in an orthogonal Cartesian coordinate system, where the quadrant of positive coordinates is the "south-west region". Therefore, on Figure 1 the origin of this coordinate system is the right-top-most corner of the *parallelogram*, where this parallelogram is defined as the points of coordinates (i, j) satisfying $0 \leq i \leq n$ and $0 \leq j \leq m$.

[2] http://en.wikipedia.org/wiki/Multiplication_algorithm#Long_multiplication

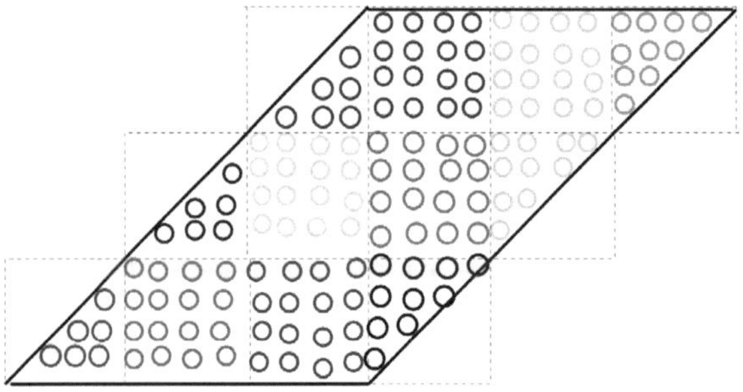

Figure 1. Dividing the work of coefficient multiplication among threadblocks.

Let r, t be two positive integers greater than 1. For simplicity let us assume that $m \geq r$ and $n \geq t$ hold. Moreover, we assume that r divides $m + 1$ and t divides $n + 1$.

The mapping between the parallelogram and the thread blocks is precisely defined as follows.

(i) We partition the parallelogram horizontally into $(m + 1)/r$ parallelograms of height r (and basis $n + 1$) that we call *long parallelograms*. Moreover, we attach to the ℓ-th long parallelogram a vector V_ℓ of length $n + r$, that we call the ℓ-th *long vector*.

(ii) We partition each long parallelogram vertically into $(n+1)/t$ equally spaced *small trapezoids*. More precisely, within a long parallelogram, the left-most and right-most trapezoids are triangles while all other trapezoids are rectangles. Observe that for each small trapezoid, all points with the same i-coordinate, say k, correspond to terms of degree k.

On Figure 1, we have $n = m = 11$. The original parallelogram is horizontally decomposed into $r = 3$ long parallelograms and each of those is decomposed into $t = 4$ rectangles/triangles.

Each thread block is in charge of computing all products within a given small trapezoid (rectangle or triangle). Let us assume for simplicity that t is a multiple of the number T of threads within a thread block. Define $s := t/T$.

Within a thread block of the ℓ-th long parallelogram, a thread is in charge of computing all terms of degrees $k, k + 1, \ldots, k + s - 1$, for some k, and adding the terms of degree $d \in \{k, k + 1, \ldots, k + s - 1\}$ in $V_\ell[d]$.

We now proceed with the addition phase. The vectors $V_0, V_1, \ldots, V_w$, with $w = (m+1)/r$, are added such that terms of the same degree are added together. This is done through a parallel reduction in $\Theta(\log(w))$ parallel steps.

3.2. Parallelism

We analyze the work and span of this algorithm. We start with the multiplication phase. We have $\frac{m+1}{r}\frac{n+1}{t}$ thread blocks. Each thread block has a work of $t(2r - 1)$ and a span of $s(2r - 1)$. Thus the work is $\Theta(n^2)$ while the parallelism is $\Theta(n^2/sr)$. For our GPU card, we choose $r = t = 2 * 9$ and $s \in \{4, 6\}$. Next, we proceed with the addition phase. The work is essentially is $2nw$, with $w = (m + 1)/r$ and the span is in $\Theta(\log(w))$. Thus, the overall parallelism is $O(n^2/\log(m))$.

3.3. Experimental Results

We have experimented the CUDA implementation of the plain univariate multiplication described in the previous section. We use both *balanced* and *unbalanced* pairs of polynomials, see Table 1 and 2 respectively. By balanced, following [21], we mean a pair of univariate polynomials of equal degree, which is a favorable case for optimized FFT-based polynomial multiplication.

degree	GPU Plain multiplication	GPU FFT-based multiplication
2^{10}	0.00049	0.0044136
2^{11}	0.0009	0.004642912
2^{12}	0.0032	0.00543696
2^{13}	0.01	0.00543696
2^{14}	0.045	0.00709072

Table 1. Comparison between plain and FFT-based polynomial multiplications for balanced pairs ($n = m$) on CUDA.

n	m	GPU Plain multiplication
2^{10}	2^8	0.00041
2^{11}	2^8	0.0005
2^{11}	2^{10}	0.00073
2^{12}	2^8	0.00057
2^{12}	2^{10}	0.0011
2^{13}	2^8	0.00074
2^{13}	2^{10}	0.0018
2^{13}	2^{12}	0.0061
2^{14}	2^8	0.0010
2^{14}	2^{10}	0.0031
2^{14}	2^{12}	0.011
2^{14}	2^{13}	0.02

Table 2. Computation time for plain multiplication on CUDA for unbalance pairs ($n \neq m$).

In Table 1, we compare the computation time of our CUDA based implementation of parallel plain multiplication with the highly optimized FFT-based multiplication in CUDA reported in [19]. Our implementation outperforms that of FFT-based multiplication until the degree 2^{12}.

FFT-based multiplication may not perform well for unbalanced pairs, see [21] for details. In plain multiplication, this is not true. Computation times for plain multiplication on CUDA of unbalanced pairs are reported in Table 2.

4. Plain Division on the GPU

Consider again two univariate polynomials over a finite field

$$a = a_n x^n + \cdots + a_1 x + a_0 \text{ and } b = b_m x^m + \cdots + b_1 x + b_0, \text{ with } n \geq m. \quad (2)$$

The only opportunity for concurrent execution is within each division step. With the above notations, the first division step computes

$$a' \leftarrow a - \frac{a_n}{b_m} x^{n-m} b, \quad (3)$$

which can be viewed as a *Gaussian elimination step*. Assuming that this is done by several thread blocks, the next division step requires to broadcast the leading coefficient of the intermediate remainder a' to all thread blocks, which is a severe performance bottleneck.

Our solution consists of letting each thread block compute the leading coefficients of s consecutive intermediate remainders, for a well chosen integer s. In this way, each thread block computes a coefficient segment (of size $2\,s$) of s consecutive intermediate remainders without synchronization. Though this increases the total work by (at most) a $\frac{1}{3}$ factor, this improves performances significantly.

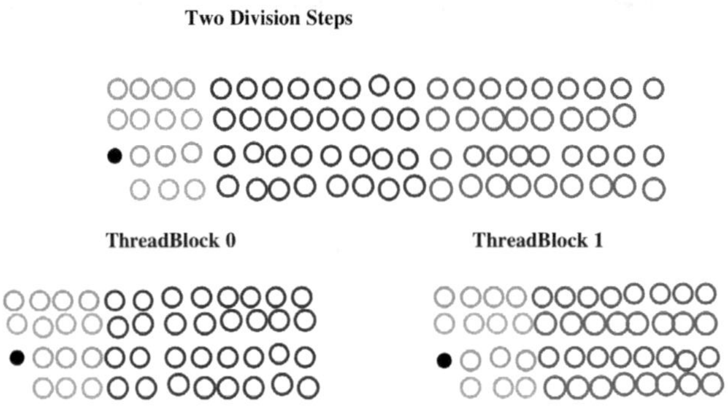

Figure 2. Each threadblock will work with a^+ (resp. b^+) along with a segment of coefficients of length $2s$.

4.1. Implementation

Recall that s is a fixed positive integer. Let a^+ (resp. b^+) be the sequence of the s *most significant* terms (that is, non-zero terms of largest degrees) of both polynomials a and b. Each thread block loads a^+ and b^+ in shared memory such that it can compute s consecutive coefficients of the quotient in the division of a by b, without synchronizing with other thread blocks.

The other coefficients of the polynomial a are divided into $(n - s)/(2s)$ segments (of consecutive, zero or non-zero, coefficients). For simplicity, let assume that $2s$ divides $n - s$.

Each thread block reads a segment of $2s$ coefficients of a and is responsible for updating these coefficients after each division step during the execution of the thread block. As each thread block can perform s division steps independently, it will also require a segment of $3s$ coefficients from the polynomial b. This is, indeed, $3s$ (and not $2s$) which is required from b since after division step the degree of the current remainder, namely a, reduces at least by 1. Overall, each thread block loads $7s$ coefficients in the shared memory.

With today's GPU cards, the maximum number of threads in a thread block is roughly one of order of magnitude away from the number of machine words that can be stored in shared memory. So we keep s in the order of the maximum number of threads in a thread block supported by a typical GPU card. As a consequence, each thread is responsible for updating $O(1)$ number of coefficients of a.

On Figure 2, $s = 4$. ThreadBlock0 is responsible for updating the a^+ part along with other 8 consecutive coefficients. In the same way, ThreadBlock1 is responsible for updating the a^+ part along with other 8 consecutive coefficients.

4.2. Parallelism

In the worst case, we need $n - m + 1$ division steps. Since one kernel call will reduce the degree of the current remainder at least by s, we have $(n - m + 1)/s$ kernel calls at most. During one kernel call, one thread is doing $O(s)$ operations. So the span of the whole procedure can be bounded by $O(n-m)$ or $O(n)$. The work is $O(m(n-m))$ or $O(nm)$. So the parallelism is $O(m)$.

4.3. Experimental Results

We have compared the running time of our CUDA implementation of the plain division with the serial C implementation of the fast division (see Chapter 9 in [10]) from the NTL [25] library. The latter algorithm is based on FFT techniques and its work fits within $O(d\log(d)\log(\log(d)))$ coefficient operations, with $d = \max(m, n)$. The input polynomials used in our experimentation are dense random (univariate) polynomials with coefficients in a finite field whose characteristic

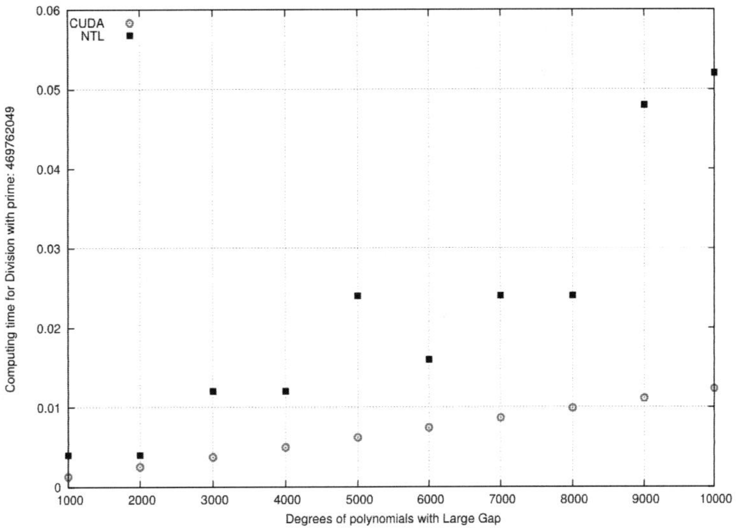

Figure 3. Comparison between parallel plain division on CUDA and fast division in NTL for univariate polynomials with large degree gap.

is a machine word prime. We use the following primes: 7, 9001 and 469762049. Our GPU code does not depend on the prime while NTL uses different algorithms depending on the prime. For a given degree pattern, the NTL running time varies at most by a factor of 2 from one of our primes to another. The degrees of our input polynomials satisfy $n = 2m$. The running time of our CUDA code outperforms that of NTL by a factor from 3 to 5, for $1,000 \leq n \leq 10,000$, see Figure 3.

5. The Euclidean Algorithm on the GPU

Recall that a and b designate two univariate polynomials over a finite field, such that either b is zero or the degree of b is not greater than that of a. The Euclidean Algorithm computes the GCD of (a, b) in the following way.

(i) if $b = 0$, then return a,

(ii) if $b \neq 0$, then return the GCD of (b, r) where r is the remainder in the division of a by b.

As for the plain division, we have no other choices than parallelizing each division step. Moreover, in order to minimize data transfer we will again let each thread block work on several consecutive division steps without synchronizing. For this to be possible, each thread block within a kernel computes the *most significant* terms of the current pair *dividend-divisor*.

5.1. Implementation

Let $s > 1$ be an integer. Each kernel call replaces the polynomial pair (a, b) by a *GCD preserving pair* (a', b') (that is, a a pair of polynomials with the same GCD as (a, b)) such that we have

$$\max(\deg(a), \deg(b)) - \max(\deg(a'), \deg(b')) \geq s.$$

Moreover, each kernel call performs at most s division steps.

- If initially $\mid \deg(a) - \deg(b) \mid \geq s$ holds, we simply use our kernel for s division steps with a fixed divisor.

- If initially $\mid \deg(a) - \deg(b) \mid < s$ holds, we modify our division kernel as follows: each thread block reads the s most significant terms for both a and b and; in addition, it reads a segment of $3s$ coefficients for both a and b. Indeed, during such a kernel call, the roles of a and b, as dividend or divisor, can be exchanged.

5.2. Parallelism

We first consider the work of the algorithm sketched above. Since $n \geq m$, the number of kernel calls is at most $\lceil \frac{n}{s} \rceil$. The number of thread blocks per kernel call is at most $\lceil \frac{n}{2s} \rceil$. The number of arithmetic operations per thread block is at most $6s^2$. Thus the work is in $O(n^2)$, as expected. However, there is an increase of work w.r.t. a serial GCD computation by a constant factor in order to reduce the amount of synchronization. Moreover, there is an increase of memory consumption w.r.t. the GPU division computation by a constant factor due to the case where the degree gap between a and b is less than s. Now, we consider the span. Since the number of kernel calls is at most $\lceil \frac{n}{s} \rceil$, and since the number of division steps per kernel call is at most s, the span is in $O(n)$, which as good as in the case of systolic arrays [2].

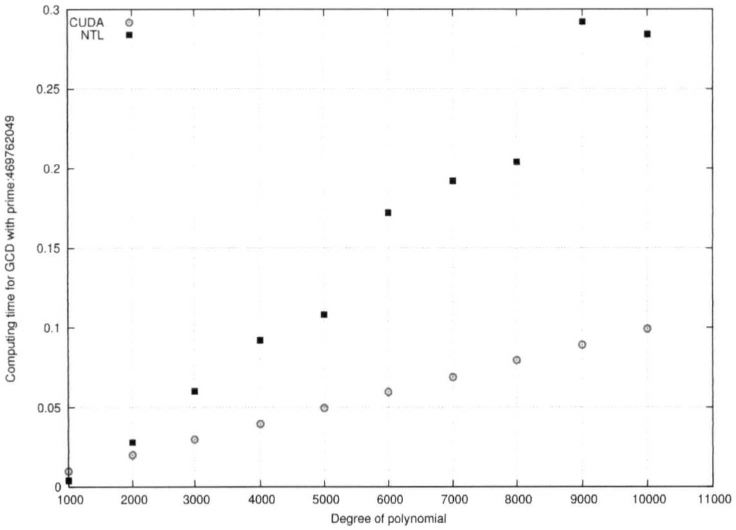

Figure 4. Comparison between parallel GCD on CUDA and FFT-based GCD in NTL for univariate polynomials, with the same degree ($n = m$).

5.3. Experimental Results

We have compared the running time of our CUDA implementation of the Euclidean Algorithm with the serial C implementation of the Half-GCD algorithm (see Chapter 11 in [10]) from the NTL [25] library. The latter algorithm is based on FFT techniques and its work fits within $O(d\log(d)\log(\log(d)))$ coefficient operations, whale that of the former algorithm amounts to $O(d^2)$ coefficient operations, for input polynomials of degree d.

As for the experimentation with the plain division, the input polynomials used in our experimentation are dense random (univariate) polynomials with coefficients in a finite field whose characteristic is a machine word prime. Here again, we use the primes 7, 9001, 469762049 and we observe that our GPU code does not depend on the prime while NTL uses different algorithms depending on the prime. For a given degree pattern, the NTL running time varies at most by a factor of 2 from one of our primes to another.

n	m	GCD on CUDA with s = 512	GCD on CUDA with s = 0
1000	500	0.010	0.024
2000	1500	0.024	0.058
3000	2500	0.039	0.108
4000	3500	0.053	0.158
5000	4500	0.069	0.203
6000	5000	0.056	0.235
7000	6000	0.066	0.282
8000	7000	0.076	0.324
9000	8000	0.087	0.367
10000	9000	0.097	0.411

Table 3. GCD implementation on CUDA with two different values of s.

Figure 4 correspond to 469762049. Indeed, we are interested in large primes since they support modular methods for polynomial system solving [6].

As reported by Figure 4, our implementation is almost three times faster than that of NTL for polynomials whose degrees range between 1,000 and 10,000. Recall that this degree range is also what is of interest for the same purpose of polynomial system solving.

The technique of computing s leading coefficients in every thread block, implemented in both division and GCD algorithm, has two major advantages. First, it reduces the number of kernel calls by a factor of s. Second, it reduces the amount of memory transfer between the global and shared memory by a factor of s.

In Table 3, we compare the computation time between two versions of our CUDA implementation of the Euclidean Algorithm. The first one sets $s = 512$ and the other one does not use at all this technique of computing s leading coefficients in every thread block. In this latter implementation, leading coefficients are kept up-to-date in the global memory such that they can be accessed by every thread block. Thus, in this scheme, every thread block works on a single division step between two updates of the leading coefficient in the global memory.

As mentioned above, the former implementation increases the work but reduces parallelization overheads in a significant manner. Table 3, shows that the former method outperforms the latter by a speedup factor varying from 2 to 4.

6. Conclusion

Motivated by the implementation of polynomial system solvers over finite fields, we were lead to parallelize plain univariate polynomial arithmetic on GPUs. For the degree range that we are targeting, namely from 2^{10} up to 2^{12}, our GPU code for plain multiplication outperforms our GPU code for FFT-based multiplication. For the degree range $2^{10} \cdots 2^{18}$, our GPU code for computing polynomial GCDs via the Euclidean Algorithm runs in linear time w.r.t the maximum degree of the input polynomials. Such sizes are sufficient for many applications.

We observed that controling parallelization overheads (synchronization on data via global memory, number of kernel calls, etc.) was essential for reaching peak performance in our implementation.

References

[1] R. A. Arce-Nazario, E. Orozco and D. Bollman. Reconfigurable hardware implementation of a multivariate polynomial interpolation algorithm. *Int. J. Reconfig. Comput.*, vol. 2010, pages 2:1–2:14, 2010.

[2] R. P. Brent, H. T. Kung, and F. T. Luk. Some Linear-time Algorithms for Systolic Arrays. In *IFIP Congress*, pages 865–876, 1983.

[3] M. F. I. Chowdhury, M. Moreno Maza, W. Pan, and É. Schost. Complexity and Performance Results for non FFT-based Univariate Polynomial Multiplication. In *Proc. of Advances in mathematical and computational methods: addressing modern of science, technology, and society*, AIP conference proceedings, volume 1368, pp. 259-262, 2011.

[4] T. H. Cormen, C. E. Leiserson, R. L. Rivest and Clifford Stein, *Introduction to Algorithms (3. ed.)* MIT Press, 2009.

[5] T. H. Cormen, C. E. Leiserson, R. L. Rivest and Clifford Stein, Introduction to Algorithms, Second Edition. The MIT Press and McGraw-Hill Book Company, 2001.

[6] X. Dahan, M. Moreno Maza, É. Schost, W. Wu, and Y. Xie. Lifting techniques for triangular decompositions. In *ISSAC'05*, pages 108–115. ACM Press, 2005.

[7] M. Frigo and S. G. Johnson. The Design and Implementation of FFTW3. In Proceedings of the IEEE Special issue on *Program Generation, Optimization, and Platform Adaptation*, volume 93, number 2, pages 216–231, 2005.

[8] M. Frigo, C. E. Leiserson, H. Prokop, and S. Ramachandran. Cache-oblivious Algorithms. In *Proceedings of the 40th Annual Symposium on Foundations of Computer Science*, FOCS '99, pages 285–297, New York, USA, October 1999.

[9] M. Frigo, C. E. Leiserson, and K. H. Randall. The Implementation of the Cilk-5 Multithreaded Language. In *ACM SIGPLAN*, 1998.

[10] J. Gathen and J. Gerhard. *Modern Computer Algebra.* Cambridge University Press, 1999.

[11] J. von zur Gathen and J. Shokrollahi. Efficient FPGA-based karatsuba multipliers for polynomials over F_2. In *Proceedings of the 12th international conference on Selected Areas in Cryptography*, SAC'05, pages 359–369, Springer-Verlag, 2006.

[12] S. A. Haque and M. Moreno Maza. *Determinant Computation on the GPU using the Condensation Method* , volume 341. J. of Physics: Conference Series, 2012.

[13] M. A. Hasan and V. K. Bhargava. Bit-Serial Systolic Divider and Multiplier for Finite Fields GF(2^m). *IEEE Trans. Comput.*, vol. 41, num. 8, pages 972–980, IEEE Computer Society, 1992.

[14] J. Hong and H. T. Kung. I/O Complexity: The Red-blue Pebble Game. In *STOC*, pages 326–333, 1981.

[15] D. E. Knuth. *The Art of Computer Programming, Volume I: Fundamental Algorithms, 3rd Edition.* Addison-Wesley Professional, 1997.

[16] D. E. Knuth. *The Art of Computer Programming, Volume II: Seminumerical Algorithms, 3rd Edition.* Addison-Wesley Professional, 1997.

[17] D. E. Knuth. *The Art of Computer Programming, Volume III: Sorting and Searching 2nd Edition.* Addison-Wesley Professional, 1998.

[18] H. T. Kung and C. L. Leiserson. Algorithms for VLSI Processor Arrays. In *Introduction to VLSI Systems*, ed. C. Mead and L. Conway, Addison-Wesley, Reading, MA, 1980, pp. 271-292.

[19] M. Moreno Maza and W. Pan. Fast polynomial arithmetic on a GPU. In *J. of Physics: Conference Series*, volume 256, 2010.

[20] M. Moreno Maza and W. Pan. Solving Bivariate Polynomial Systems on a GPU. In *J. of Physics: Conference Series*, volume 341, 2011.

[21] M. Moreno Maza and Y. Xie. Balanced dense polynomial multiplication on multicores. In *Int. J. Found. Comput. Sci.*, volume 22, numbe 5, pages 1035-1055, 2011.

[22] M. Moreno Maza and Y. Xie. FFT-based dense polynomial arithmetic on multi-cores. In D. Mewhort, editor, *Proc. HPCS 2009*, volume 5976 of *LNCS*, Heidelberg, 2010. Springer-Verlag Berlin.

[23] M. Nele, S. B. Örs, B, Preneel and J. Vandewalle An FPGA Implementation of a Montgomery Multiplier Over GF(2^m). *J. of Computers and Artificial Intelligence*, vol. 23, num. 5, pages 487-499, 2004.

[24] M. Püschel, J. M. F. Moura, J. Johnson, D. Padua, M. Veloso, B. Singer, J. Xiong, F. Franchetti, A. Gacic, Y. Voronenko, K. Chen, R. W. Johnson and N. Rizzolo. SPIRAL: Code Generation for DSP Transforms, In Proceedings of the IEEE, special issue on *Program Generation, Optimization, and Adaptation*, volume 93, numbe 2, pages 232-275, 2005.

[25] V. Shoup. NTL: A Library for doing Number Theory. `http://www.shoup.net/ntl/`

[26] C. Yap. *Fundamental Problems in Algorithmic Algebra.* Princeton University Press, 1993.

High Performance Computing Symposium 2012 (HPCS2012) IOP Publishing
Journal of Physics: Conference Series **385** (2012) 012015 doi:10.1088/1742-6596/385/1/012015

On the Factor Refinement Principle and its Implementation on Multicore Architectures

Md. Mohsin Ali, Marc Moreno Maza and Yuzhen Xie

University of Western Ontario, London, Ontario, Canada

E-mail: md.ali@anu.edu.au,moreno@csd.uwo.ca,yxie@maplesoft.com

Abstract.
We propose a divide and conquer adaptation of the factor refinement algorithm of Bach, Driscoll and Shallit. For an ideal cache of Z words, with L words per block, the original approach suffers from $O(n^2/L)$ cache misses, meanwhile our adaptation incurs $O(n^2/ZL)$ cache misses only. We have realized a multithreaded implementation of the latter using `Cilk++` targeting multicores. Our code achieves linear speedup on 16 cores for sufficiently large input data.

1. Introduction

The implementation of non-linear polynomial system solvers is a very active research area. It has been stimulated during the past ten years by two main progresses. Firstly, methods for solving such systems have been improved by the use of so-called modular techniques and asymptotically fast polynomial arithmetic. See the landmark textbook [11] for this research area. Secondly, the democratization of supercomputing, thanks to hardware acceleration technologies (multicores, general purpose graphics processing units) creates the opportunity to tackle harder problems.

One central issue in the implementation of polynomial system solvers is the elimination of redundant expressions that frequently occur during intermediate computations, with both numerical methods and computer algebra methods. Factor refinement, also known as coprime factorization, is a popular technique for removing repeated factors within a set of polynomials. Coprime factorization has numerous applications in number theory and polynomial algebra, see the papers [5, 6] and the web site `http://cr.yp.to/coprimes.html` by Bernstein.

Algorithms for coprime factorization have generally been designed with algebraic complexity as the complexity measure to optimize. However, with the evolution of computer architecture, parallelism and data locality are becoming major measures of performance, in addition to the traditional ones, namely serial running time and allocated space.

In this paper, we revisit the factor refinement algorithms of Bach, Driscoll and Shallit [4] based on quadratic arithmetic. We show that their *augment refinement principle* leads to a highly efficient algorithm in terms of parallelism and data locality. Our approach is inspired by the paper "Parallel Computation of the Minimal Elements of a Poset" [8]. Figure 1 illustrates the divide and conquer scheme of our algorithm. Once the original problem has been divided into sufficiently small sub-problems, those sub-problems are solved serially and then their solutions are merged in parallel. Here's the difficult point. As demonstrated in [3], a straightforward implementation of the merge phase can bring a bottleneck in terms of *cache complexity*; see [9, 12] for this notion. The main contribution of this paper is to overcome this difficulty and exhibit a factor refinement algorithm which is efficient in terms of cache complexity and parallelism.

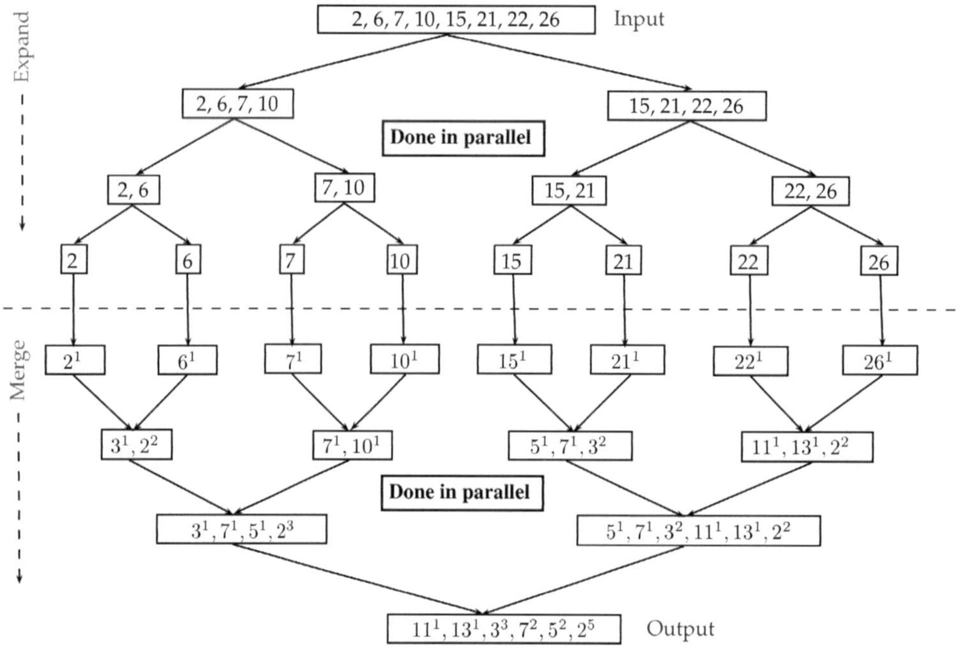

Figure 1. Example of algorithm execution.

On input data of size n, under the condition that each arithmetic operation (integer division and integer GCD computation) has a unit cost, this parallel algorithm features $O(n^2)$ work, $O(Cn)$ span, and thus $O(n/C)$ parallelism, where C is the size threshold below which serial code is run. Other approaches [5, 7] have a better work, but as shown in [3] they are not suitable for implementation on multicore architectures due to high parallelization overheads. This motivates the work reported in this paper.

For an ideal cache of Z words, with L words per cache-line, with C small enough, for any input of size n, the number of cache misses of our algorithm is

$$Q(n) = O(n^2/ZL + n^2/Z^2),$$

under the assumption that each input or output sequence is *packed*, and each sequence item is stored in *one machine word*. Moreover, under the tall cache assumption, the above estimate further simplifies to $Q(n) = O(n^2/ZL)$. All these assumptions hold in our implementation for the case of univariate polynomials over a small prime field. Moreover, our code enforces *balancing of sub-problems* for optimizing the share of computer resources among threads.

Our theoretical results are confirmed by an experimentation, based on a multicore implementation in the `Cilk++` concurrency platform [10, 13]. For problems on integers, we use the GMP library [1] while for problems on polynomials, we use the "Basic Polynomial Algebra Subroutines (BPAS)" library [14].

2. The Factor Refinement Principle

We review below the notions of factor refinement, coprime factorization and GCD-free basis. Let $n_1, \ldots, n_s$ be all integers or all univariate polynomials over a field $\mathbb{K}$. We say that $n_1, \ldots, n_s$ form a *GCD-free basis* whenever $\gcd(n_i, n_j) = 1$ for all $1 \leq i < j \leq s$. Let $m_1, \ldots, m_r$ be other elements of the same type as the n_i's and let m be the product of the m_j's. Let $e_1, \ldots, e_s$ be positive integers. We say that the pairs $(n_1, e_1), \ldots, (n_s, e_s)$ form a *factor refinement* of $m_1, \ldots, m_r$ if the following conditions hold:

(*i*) $n_1, n_2, \ldots, n_s$ form a GCD-free basis,

(*ii*) for every $1 \leq i \leq r$ there exists non-negative integers $f_1, \ldots, f_s$ such that we have $\prod_{1 \leq j \leq s} n_j^{f_j} = m_i$,

(*iii*) $\prod_{1 \leq i \leq s} n_i^{e_i} = m$.

When this holds, we also say that $(n_1, e_1), \ldots, (n_s, e_s)$ is a *coprime factorization* of m. For instance, $5^1, 6^2, 7^1$ is a refinement of 30 and 42 while $5^1, 6^2, 7^1$ is a coprime factorization of 1260.

Given a partial factorization of an integer m, say $m = m_1 m_2$, one can compute $d = \gcd(m_1, m_2)$ and write

$$m = (m_1/d)(d^2)(m_2/d).$$

If this process is continued until all the factors of m are pairwise coprime, one obtains a coprime factorization of m within $O(\text{size}(m)^3)$ bit operations. In their landmark 1988 paper [4], Bach, Driscoll and Shallit observed that, by keeping track of the pairs (n_j, n_k) in an *ordered pair list* such that only elements adjacent in the list can have a nontrivial GCD, one computes a coprime factorization of m within $O(\text{size}(m)^2)$ bit operations. In their proof, they simply rely on plain (or quadratic) arithmetic. Using asymptotically fast algorithms for the case of univariate polynomials, D. Bernstein [5] on one hand and, Dahan, Moreno Maza, Schost, Xie [7] on another, have obtained an estimate of $O(d \log_2^4(d) \log_2(\log_2(d)))$, where d is the sum of the degrees of the input polynomials. We note that none of these previous works has considered the question of cache complexity for factor refinement or, equivalently for coprime factorization or GCD-free basis computation.

In the sequel of this paper, we focus on the case of univariate polynomials, for which an additional notion is essential. A univariate polynomial f with coefficients in a field $\mathbb{K}$ is said *squarefree* if for every non-constant polynomial $g \in \mathbb{K}[x]$, the polynomial g^2 does not divide f. Note that it is possible to compute a factor refinement $(f_1, e_1), \ldots, (f_s, e_s)$ of f such that the $f_1, \ldots, f_s$ are pairwise coprime squarefree polynomials and $e_i < e_j$ holds for all $1 \leq i < j \leq p$. This factor refinement is called a *squarefree factorization* of f; see [11, 15] for details.

3. A Divide and Conquer Approach

Algorithm 5 presented in this section is an efficient algorithmic solution in terms of data locality and parallelism for coprime factorization of integers or univariate polynomials. For an ideal cache of Z words, with L words per cache-line, for an input data of size of n, under the assumptions stated in the introduction, Algorithm 5 incurs $O(n^2/ZL)$ cache misses. In contrast, it is not hard to check, see [3], that the original algorithm by Bach, Driscoll and Shallit (based on ordered pair lists) suffers from $O(n^2/L)$ cache misses. Modern L1 caches typically match $L = 2^9$ and $Z = 2^{16}$ while for L3 caches, one can have $Z = 2^{26}$ or more.

This substantial gain of Algorithm 5 is obtained thanks to a cache efficient procedure (Algorithm 4) for merging two coprime factorizations. Before analyzing Algorithm 5, we describe its specifications and those of its subroutines, namely Algorithms 1, 2, 3, and 4:

- Algorithm 1 receives two square-free polynomials $a, b \in \mathbb{K}[x]$ and two positive integers e, f. Its output is a factor refinement of a^e, b^f.
- Algorithm 2 receives a square-free polynomial $a \in \mathbb{K}[x]$, a positive integer e, a sequence of square-free pairwise coprime polynomials $(b_1, b_2, \ldots, b_n)$ of $\mathbb{K}[x]$ and a sequence of positive integers $(f_1, f_2, \ldots, f_n)$. The output is a factor refinement of $a^e, {b_1}^{f_1}, \ldots, {b_n}^{f_n}$.

Algorithm 1: `PolyRefine`(a, e, b, f)

Input: $a, b \in \mathbb{K}[x]$ squarefree polynomials over a field $\mathbb{K}$ and e, f two positive integers.
Output: (c, u, G, V, d, w) where $c, d \in \mathbb{K}[x]$ and $u, w \in \mathbb{N}$ and G is a sequence $(g_1, \ldots, g_s)$ of polynomials of $\mathbb{K}[x]$ and V is a sequence $(v_1, \ldots, v_s)$ of positive integers such that $(c, u), (g_1, v_1), \ldots, (g_s, v_s), (d, w)$ is a factor refinement of $(a, e), (b, f)$.

$g \leftarrow \gcd(a, b)$; $a' \leftarrow a$ quotient g; $b' \leftarrow b$ quotient g;
if $g = 1$ **then**
    return $(a, e, \varnothing, \varnothing, b, f)$; `// Here` $\varnothing$ `designates the empty sequence`
else if $a = b$ **then**
    return $(1, 1, (a), (e + f), 1, 1)$
else
    $(\ell_1, e_1, G_1, V_1, r_1, f_1) \leftarrow$ `PolyRefine`$(a', e, g, e + f)$;
    $(\ell_2, e_2, G_2, V_2, r_2, f_2) \leftarrow$ `PolyRefine`(r_1, f_1, b', f);
    if $\ell_2 \neq 1$ **then**
        $G_2 \leftarrow G_2 + (\ell_2)$; `// Here + designates sequence concatenation`
        $V_2 \leftarrow V_2 + (e_2)$;
    return $(\ell_1, e_1, G_1 + G_2, V_1 + V_2, r_2, f_2)$;

Algorithm 2: `MergeRefinePolySeq`(a, e, B, F)

Input: A square-free polynomial $a \in \mathbb{K}[x]$, a positive integer e, a sequence of square-free pairwise coprime polynomials $(b_1, b_2, \ldots, b_n)$ of $\mathbb{K}[x]$ and a sequence of positive integers $(f_1, f_2, \ldots, f_n)$.
Output: (ℓ, m, Q, R, S, T) where $\ell \in \mathbb{K}[x]$, $m \in \mathbb{N}$, $Q = (q_1, \ldots, q_s)$ and $S = (s_1, \ldots, s_p)$ are 2 polynomial sequences of $\mathbb{K}[x]$, $R = (r_1, \ldots, r_s)$ and $T = (t_1, \ldots, t_p)$ are 2 positive integer sequences s. t. $(\ell, m), (q_1, r_1), \ldots, (q_s, r_s), (s_1, t_1), \ldots, (s_p, t_p)$ is a factor refinement of $a^e, {b_1}^{f_1}, \ldots, {b_n}^{f_n}$.

$\ell_0 \leftarrow a$; $m_0 \leftarrow e$; $Q \leftarrow \varnothing$; $R \leftarrow \varnothing$; $S \leftarrow \varnothing$; $T \leftarrow \varnothing$;
for i from 1 to n **do**
    $(\ell_i, m_i, G_i, V_i, d_i, w_i) \leftarrow$ `PolyRefine`$(\ell_{i-1}, m_{i-1}, b_i, f_i)$;
    $Q \leftarrow Q + G_i$;
    $R \leftarrow R + V_i$;
    if $d_i \neq 1$ **then**
        $S \leftarrow S + (d_i)$;
        $T \leftarrow T + (w_i)$;
return $(\ell_n, m_n, Q, R, S, T)$;

- Algorithm 3 takes two sequences of square-free pairwise coprime polynomials $(a_1, \ldots, a_n)$ and $(b_1, b_2, \ldots, b_r)$ of $\mathbb{K}[x]$ together with two sequences of positive integers $(e_1, e_2, \ldots, e_n)$ and $(f_1, f_2, \ldots, f_r)$. The output is a factor refinement of ${a_1}^{e_1}, \ldots, {a_n}^{e_n}, {b_1}^{f_1}, \ldots, {b_r}^{f_r}$.
- The specifications of Algorithm 4 are the same as those of Algorithm 3. The difference is that Algorithm 4 runs in parallel and uses Algorithm 3 as serial base case.
- Finally,Algorithm 5 takes a sequence of square-free polynomials $(m_1, m_2, \ldots, m_k) \in \mathbb{K}[x]$ as input and generates as output a sequence of univariate polynomials $(n_1, n_2, \ldots, n_s) \in \mathbb{K}[x]$

Algorithm 3: `MergeRefineTwoSeq`(A, E, B, F)

Input: Two sequences of square-free pairwise coprime polynomials $A = (a_1, \ldots, a_n)$ and $B = (b_1, b_2, \ldots, b_r)$ of $\mathbb{K}[x]$ together with two sequences of positive integers $E = (e_1, e_2, \ldots, e_n)$ and $F = (f_1, f_2, \ldots, f_r)$.

Output: (L, M, Q, R, S, T) where $L = (\ell_1, \ldots, \ell_h)$, $Q = (q_1, \ldots, q_s)$ and $S = (s_1, \ldots, s_p)$ are three sequences of polynomials of $\mathbb{K}[x]$, $M = (m_1, \ldots, m_h)$, $R = (r_1, \ldots, r_s)$ and $T = (t_1, \ldots, t_p)$ are three sequences of positive integers such that $(\ell_1, m_1), \ldots, (\ell_h, m_h), (q_1, r_1), \ldots, (q_s, r_s), (s_1, t_1), \ldots, (s_p, t_p)$ is a factor refinement of $(a_1, e_1), \ldots, (a_n, e_n), (b_1, f_1), \ldots, (b_r, f_r)$.

```
L ← ∅; M ← ∅; Q ← ∅; R ← ∅; S_0 ← B; T_0 ← F;
for i from 1 to n do
    (ℓ_i, m_i, Q_i, R_i, S_i, T_i) ← MergeRefinePolySeq(a_i, e_i, S_{i-1}, T_{i-1}) ;
    Q ← Q + Q_i ;
    R ← R + R_i ;
    if ℓ_i ≠ 1 then
        L ← L + (ℓ_i) ;
        M ← M + (m_i) ;
return (L, M, Q, R, S_n, T_n);
```

Algorithm 4: `MergeRefinementDNC`(A, E, B, F)

Input: Two sequences $A = (a_1, \ldots, a_n)$ and $B = (b_1, b_2, \ldots, b_r)$ of square-free pairwise coprime polynomials in $\mathbb{K}[x]$ together with two sequences of positive integers $E = (e_1, e_2, \ldots, e_n)$ and $F = (f_1, f_2, \ldots, f_r)$.

Output: (L, M, Q, R, S, T) where $L = (\ell_1, \ldots, \ell_h)$, $Q = (q_1, \ldots, q_s)$ and $S = (s_1, \ldots, s_p)$ are three sequences of polynomials of $\mathbb{K}[x]$, $M = (m_1, \ldots, m_h)$, $R = (r_1, \ldots, r_s)$ and $T = (t_1, \ldots, t_p)$ are three sequences of positive integers such that $(\ell_1, m_1), \ldots, (\ell_h, m_h), (q_1, r_1), \ldots, (q_s, r_s), (s_1, t_1), \ldots, (s_p, t_p)$ is a factor refinement of $(a_1, e_1), \ldots, (a_n, e_n), (b_1, f_1), \ldots, (b_r, f_r)$.

```
if n ≤ BASESIZE or r ≤ BASESIZE then
    return MergeRefineTwoSeq(A, E, B, F);
else
    Divide A, E, B, and F into two halves called A_1, A_2, E_1, E_2, B_1, B_2, and F_1, F_2,
    respectively ;
    (L_1, M_1, Q_1, R_1, S_1, T_1) ← spawn MergeRefinementDNC(A_1, E_1, B_1, F_1) ;
    (L_2, M_2, Q_2, R_2, S_2, T_2) ← spawn MergeRefinementDNC(A_2, E_2, B_2, F_2) ;
    sync ;
    (L_3, M_3, Q_3, R_3, S_3, T_3) ← spawn MergeRefinementDNC(L_1, M_1, S_2, T_2) ;
    (L_4, M_4, Q_4, R_4, S_4, T_4) ← spawn MergeRefinementDNC(L_2, M_2, S_1, T_1) ;
    sync ;
    return {L_3 + L_4, M_3 + M_4, Q_1 + Q_2 + Q_3 + Q_4, R_1 + R_2 + R_3 + R_4, S_3 + S_4, T_3 + T_4} ;
```

Algorithm 5: `ParallelFactorRefinementDNC`(A)

Input: A sequence $A = (m_1, m_2, \ldots, m_k)$ of square-free polynomials of $\mathbb{K}[x]$.
Output: A sequence of square-free pairwise coprime polynomials $N = (n_1, n_2, \ldots, n_s) \in \mathbb{K}[x]$, and a sequence of positive integers $E = (e_1, e_2, \ldots, e_s)$ such that $(n_1, e_1), (n_2, e_2), \ldots, (n_s, e_s)$ is a factor refinement of $m_1, m_2, \cdots, m_k$.

if $k < 2$ **then**
  return $(m_1), (1)$;
else
  Divide A into two subsequences called A_1 and A_2;
  $(X_1, Y_1) \leftarrow$ **spawn** `ParallelFactorRefinementDNC`(A_1) ;
  $(X_2, Y_2) \leftarrow$ **spawn** `ParallelFactorRefinementDNC`(A_2) ;
  sync;
  return `MergeRefinementDNC`(X_1, Y_1, X_2, Y_2);

and a sequence of positive integers $(e_1, e_2, \ldots, e_s)$ such that $((n_1, e_1), (n_2, e_2), \ldots, (n_s, e_s))$ is a refinement of $(m_1, m_2, \cdots, m_k)$. Moreover, $n_1, n_2, \ldots, n_s$ are square-free.

Proposition 1 analyzes the parallelism of Algorithm 5 for the fork-join parallelism model [10] under one simplification hypothesis, which we state below.

We assume that each polynomial operation (division or GCD computation) involved in Algorithm 5 or its subroutines has a unit cost. Obviously, this assumption does not seem realistic. However, if D is the maximum degree of an input polynomial and if each arithmetic operation (addition, multiplication, inversion) in $\mathbb{K}$ has a constant bit cost, then each subsequent polynomial operation performed by Algorithm 5 or its subroutines runs in $O(D^2)$ bit operations. Therefore, this assumption can still be seen as a first approximation of what really happens.

In order to analyze Algorithm 5 and its subroutines, one needs to choose a measure of the input data. Our assumption suggests the following choice. For each of Algorithms 1, 2, 3, 4 or 5, the input size is the total number of polynomials in the input.

Proposition 1 *With the above hypotheses, for a size n input, the work, span, and parallelism of Algorithm 5 are respectively $O(n^2)$, $O(Cn)$ and $O(n/C)$, where C is the threshold* BASESIZE.

The proof of Proposition 1 can easily be made by adapting the proof techniques of [8], see [3].

We turn now our attention to cache complexity and analyze the cache complexity of the serial versions of our algorithms. We start by specifying how data is layed out in memory. We assume that each input or output sequence in Algorithms 1, 2, 3 or 4 is *packed*, that is, its successive polynomials occupy consecutive memory slots. We also assume that each element of the field $\mathbb{K}$ can be stored in one machine word.

We specify a few helpful notations for establishing our cache complexity results. We denote by $|p|$ the degree of a non-zero univariate polynomial p over $\mathbb{K}$. For a finite polynomial sequence $P = (p_1, \ldots, p_n)$, we denote by $|P|$ the sum of the degrees of $p_1, \ldots, p_n$.

Nest, we make a second assumption: each integer in any input or output sequence of positive integers is stored in one machine word; moreover, any polynomial in any input or output sequence of Algorithms 1, 2, 3 or 4 is non-constant and monic[1]. Consequently, any polynomial p in any input or output sequence of Algorithms 1, 2, 3 or 4 can be stored within $|p|$ machine words.

[1] As they are currently stated, Algorithms 1, 2, 3 or 4 do not meet this latter assumption However, enforcing this assumption only requires simple transformations that we do not reproduce to keep pseudo-code easier to read.

Lemma 1 *Under the above assumptions, with the notations of Algorithm 1, we have*

$$|c| + |G| \leq |a| \quad \text{and} \quad |G| + |d| \leq |b|. \tag{1}$$

PROOF ▷ We proceed by induction on the sum of the degrees of the input polynomials of Algorithm 1, that is, $|a| + |b|$. First, we observe that, at Line 4, if $g = 1$ holds, then the conclusion trivially holds. Indeed, in this case, we have $c = a$, $G = \varnothing$ and $d = b$. Similarly, at Line 8, if $a = b$ holds, the conclusion also holds. These two cases cover, in particular, the base case of the induction, that is, $|a| + |b| = 2$. Now, we assume that g has a positive degree. Thus, we can apply the induction hypothesis to the two recursive calls at Lines 14 and 16, since we have $|a'| + |g| < |a| + |b|$ and thus $|r_1| + |b'| \leq |g| + |b'| < |a| + |b|$.

Next, we shall estimate $|\ell_1|$, $|G_1 + G_2|$ and $|r_2|$ at the last return point of Algorithm 1, that is, at Line 24. We designate by G_2^0 the value of G_2 after executing Line 16 and before executing Line 18. Then we have:

$$\begin{aligned} |\ell_1| + |G_1 + G_2| &\leq |\ell_1| + |G_1| + |G_2| && \text{by definition} \\ &\leq |\ell_1| + |G_1| + |G_2^0| + |\ell_2| && \text{by definition} \\ &\leq |a'| + |g| && \text{by induction} \\ &\leq |a| && \text{since } a = a'g. \end{aligned}$$

Similarly, we have

$$\begin{aligned} |G_1 + G_2| + |r_2| &\leq |G_1| + |G_2| + |r_2| && \text{by definition} \\ &\leq |G_1| + |\ell_2| + (|G_2^0| + |r_2|) && \text{by definition} \\ &\leq |G_1| + |r_1| + |b'| && \text{by induction} \\ &\leq |g| + |b'| && \text{by induction} \\ &\leq |b| && \text{since } b = b'g. \end{aligned}$$

This completes the proof. ◁

Lemma 2 *Under the above assumptions, with the notations of Algorithm 2, we have*

$$|\ell| + |Q| \leq |a| \quad \text{and} \quad |Q| + |S| \leq |B|. \tag{2}$$

PROOF ▷ We use the notations of the pseudo-code of Algorithm 2. We have the following inequalities:

$$\begin{aligned} |\ell_n| + |Q| &\leq |\ell_n| + |G_n| + |G_{n-1}| + \cdots + |G_1| && \text{by definition} \\ &\leq |\ell_{n-1}| + |G_{n-1}| + \cdots + |G_1| && \text{by Lemma 1} \\ &\vdots && \vdots \\ &\leq |\ell_1| + |G_1| && \text{by Lemma 1} \\ &\leq |l_0| && \text{by Lemma 1} \\ &\leq |a| && \text{by definition.} \end{aligned}$$

With Lemma 1, we also have:

$$|Q| + |S| \leq \sum_{i=1}^{i=n} (|G_i| + |d_i|) \leq \sum_{i=1}^{i=n} |b_i| \leq |B|.$$

This completes the proof. ◁

Lemma 3 *Under the above assumptions, with the notations of Algorithm 3, we have*

$$|L| + |Q| \leq |A| \quad \text{and} \quad |Q| + |S_n| \leq |B|. \tag{3}$$

The proof technique is similar to that of the previous two lemmas, so we skip it here.

Proposition 2 *Under the above assumptions, for an input of size n, each of the Algorithms 1, 2 and 3 can be run in space $\Theta(n)$ bits.*

PROOF ▷ Lemmas 1, 2 and 3 imply that for an input of size n, the output size is at most n. Recall that by input size or output size, we mean the total number of polynomial coefficients (except the leading coefficients since they are all set to 1). The output contains also sequences of positive integers, the number of those being at most n. Finally, we observe that each of the Algorithms 1, 2 and 3 does not require extra memory space other than: the space for the input and output data, and a constant number of pointers and index variables. ◁

Lemma 4 *Under the above assumptions, for an ideal cache of Z words, with L words per cache-line, there exists a positive constant α such that for an input of size n, satisfying $n < \alpha Z$, the number of cache misses of Algorithm 3 is $O(n/L + 1)$.*

PROOF ▷ By Proposition 2, there exists a positive constant β such that for an input of size n, the memory requirement for running Algorithm 3 is at most βn bits. And those βn bits, up to a constant number of them, are used for storing a number (independent of n) of sequences of polynomials and positive integers. Now recall that each sequence of polynomials (resp. positive integers) is stored in an array. Recall also that loading to cache (resp. writing back to main memory) an array of length ℓ requires $O(\ell/L + 1)$ cache misses. This conclusion follows. ◁

Theorem 1 *Under the above assumptions, for an ideal cache of Z words, with L words per cache-line, for C small enough, for a size n input, the number of cache misses of Algorithm 4 is $Q(n) = O(n^2/ZL + n^2/Z^2)$, which becomes $Q(n) \in O(n^2/ZL)$ with tall cache assumption.*

PROOF ▷ It follows from Lemma 4 and the recursive structure of Algorithm 4 that there exists a positive constant α such that $Q(n)$ satisfies the following relation:

$$Q(n) \leq \begin{cases} O(n/L+1) & \text{for } n < \alpha Z \\ 4Q(n/2) + \Theta(1) & \text{otherwise,} \end{cases}$$

provided that $C < \alpha Z$ holds. Strictly speaking the above recurrence assumes that each of the input polynomial sequences A and B can be split into two sequences of approximately the same size. When the total number of polynomials in A and B is large, this is likely. When it is small, the condition $n < \alpha Z$ is likely to hold and we are "out of the woods". A more formal treatment can be done using standard (but quite involved) proof techniques, as for the parallelization of *quick-sort* algorithm[2]. The above recurrence leads to the following inequality for all $n \geq 2$:

$$\begin{aligned} Q(n) &\leq 4Q(n/2) + \Theta(1) \\ &\leq 4[4Q(n/4) + \Theta(1)] + \Theta(1) \\ &\vdots \\ &\leq 4^k Q(n/2^k) + \sum_{j=0}^{k-1} 4^j \Theta(1) \end{aligned}$$

where $k = \lceil \log_2(n/\alpha Z) \rceil$. Since we have $n/2^k \leq \alpha Z$, we deduce:

$$\begin{aligned} Q(n) &\leq (n/\alpha Z)^2(\alpha Z/L + 1) + \Theta((n/\alpha Z)^2) \\ &= O(n^2/ZL + n^2/Z^2). \end{aligned}$$

This completes the proof. ◁

[2] For details, see the slides of the lecture *Analysis of Multithreaded Algorithms* available at `http://www.csd.uwo.ca/~moreno/CS9624-4435-1011.html`

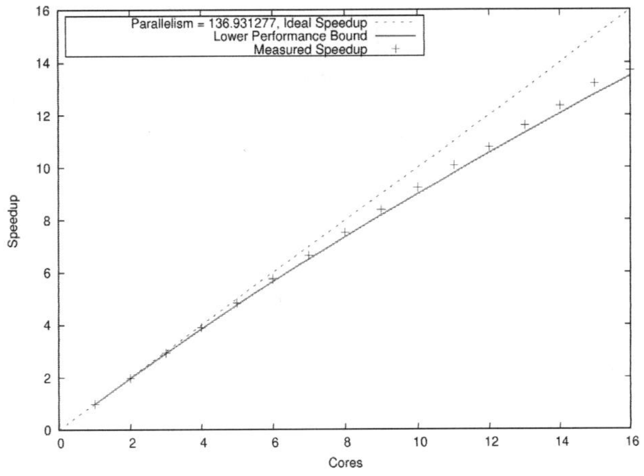

Figure 2. `Cilkview`-based scalability analysis of our parallel factor refinement algorithm for as a set of dense 5,000 univariate polynomials of degree up to 200 as input.

4. Experimentation

We have implemented our algorithm in `Cilk++` and tested it successfully with both integers and polynomials. Figure 2 generated with `Cilkview`, shows scalability results on an Intel(R) Xeon(R) (64 bit) 16-core node, with CPU (E7340) Speed 2.40GHz, 128.0 GB of RAM for an input data set of 5,000 polynomials of various degrees up to 200. Coefficients are taken modulo a machine word prime. On 16 cores, a speedup factor of 14 is reached. Figure 3 shows that similar performance results have been obtained with an input data set of 200,000 machine word integers. See [3] for more experimental results. Within MAPLE [2], we have realized serial implementations of both the original algorithm of Bach, Driscoll and Shallit [4] and our divide and conquer adaptation. For sufficiently large input, the latter outperforms the former by a factor ranging from 3 to 6.

5. Concluding remarks

We have presented a cache efficient algorithm for coprime factorization, based on the factor refinement principle of Bach, Driscoll and Shallit [4]. The high ratio *work to cache complexity* of our approach (namely $O(ZL)$ for an ideal cache of Z words, with L words per block) suggests that it is well suited for modern architectures, in particular multicores. Experimental comparison between the serial implementations of the original algorithm of [4] and ours confirm the benefits of the latter.

The divide and conquer structure of our approach leads to a multithreaded algorithm with a parallelism of $O(n)$ (in the fork-join model) for an input data of size n. Different optimization techniques (work load balancing among sub-problems, data packing) are used in our our `Cilk++` implementation which achieves linear speedup on 16 cores for sufficiently large input data.

We believe that this work illustrates the importance of cache complexity in the design of efficient algorithm targeting multicore architectures. As shown in [3], alternative algorithmic approaches for coprime factorization fail on these architectures either because they have a low ratio work to cache complexity or because of parallelization overheads.

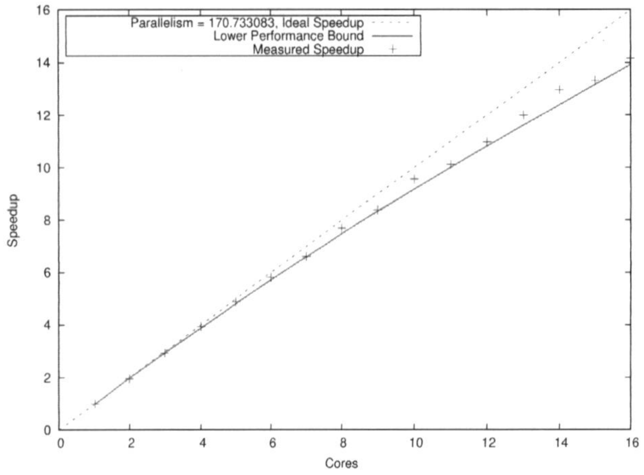

Figure 3. `Cilkview`-based scalability analysis of our parallel factor refinement algorithm for a set of 200,000 integers as input.

References

[1] GMP: Arithmetic without limitations. `http://gmplib.org`.

[2] Maplesoft, Waterloo, Ontario, Cananda. *The computer algebra system Maple*, `http://www.maplesoft.com`.

[3] Md. M. Ali *On the Factor Refinement Principle and its Implementation on Multicore Architectures* Masters Thesis, The University of Western Ontario, 2011.

[4] E. Bach, J. Driscoll, and J. Shallit. Factor Refinement. In *Symposium on Discrete Algorithms*, pages 201–211, 1990.

[5] D. J. Bernstein. Factoring into Coprimes in Essentially Linear Time. *J. Algorithms*, 54(1):1–30, 2005.

[6] D. J. Bernstein, H. W. Lenstra Jr., and J. Pila. Detecting Perfect Powers by Factoring into Coprimes. *Math. Comput.*, 76(257):385–388, 2007.

[7] X. Dahan, M. Moreno Maza, É. Schost, W. Wu, and Y. Xie. On the Complexity of the D5 Principle. *SIGSAM Bull.*, 39:97–98, September 2005.

[8] C. E. Leiserson, L. Li, M. Moreno Maza, and Y. Xie. Parallel Computation of the Minimal Elements of a Poset. In *4th International Workshop on Parallel and Symbolic Computation*, pages 53–62, 2010.

[9] M. Frigo, C. E. Leiserson, H. Prokop, and S. Ramachandran. Cache-oblivious Algorithms. In *Proc. of the 40th Annual Symposium on Foundations of Computer Science*, FOCS '99, pages 285–297, 1999.

[10] M. Frigo, C. E. Leiserson, and K. H. Randall. The Implementation of the Cilk-5 Multithreaded Language. In *ACM SIGPLAN*, 1998.

[11] J. von zur Gathen and J. Gerhard. *Modern Computer Algebra.* Cambridge University Press, 2nd ed., 2003.

[12] J. Hong and H. T. Kung. I/O Complexity: The Red-blue Pebble Game. In *STOC*, pages 326–333, 1981.

[13] C. E. Leiserson. The Cilk++ Concurrency Platform. In *DAC '09: Proceedings of the 46th Annual Design Automation Conference*, pages 522–527, New York, NY, USA, 2009. ACM.

[14] M. Moreno Maza and Y. Xie. FFT-based Dense Polynomial Arithmetic on Multi-cores. In D.J.K. Mewhort, editor, *Proc. HPCS 2009*, volume 5976 of *LNCS*, Heidelberg, 2010. Springer-Verlag Berlin.

[15] D. Y. Y. Yun. On Square-free Decomposition Algorithms. In *Proceedings of the Third ACM Symposium on Symbolic and Algebraic Computation*, SYMSAC '76, pages 26–35, New York, NY, USA, 1976. ACM.

AUTHOR INDEX

Institute of Physics
Dirac House, Temple Back
Bristol BS1 6BE UK